THE BEST
OF
NORTHWEST ARIZONA
AND THE
GRAND CANYON

BY

L.J. ETTINGER

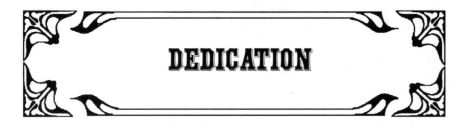

DEDICATION

This book is dedicated to Willard C. Lacy, Spencer R. Titley and Eldred D. Wilson, three of the giants of Arizona geology and mining, all of whom I had the privilege of knowing at the University of Arizona in the early 1960s.

Library of Congress Control Number: 2005905766
ISBN 0-9614840-7-1

Copyright © 2005 by L.J. Ettinger

Printed in the U.S.A.

Published by L.J. Ettinger
1991 Saddleback Road, Reno, Nevada 89521
Phone: (775) 847-9303

Previous Page:
"Guardian of the Canyon." If you use your imagination you can see the profile of the Native American on the cliffs overlooking the Grand Canyon as seen from the Bright Angel Trail. Photo by Ann Ettinger.

FOREWORD

As a geologist as well as an attorney, Len Ettinger is uniquely qualified to chronicle the history of this obscure section of the American West and this is a book deserving of a place in the collections of those interested in this rough land and in the glove compartments of the vehicles they drive. The author's topical narrative of the saga of the geological history, native peoples, the coming of the Spanish, early American explorations, the Mormons, military posts, mining, railroading, community development and much more is the book's most attractive feature, enabling the reader to pursue various aspects of the past without the distraction of other tangential elements. More histories should be put together in this fashion. Northwestern Arizona and all the history set forth here is still out there. Enjoy. Share. Tell your friends!

— Phillip I. Earl
Nevada Historical Society — Curator, Emeritus

ACKNOWLEDGEMENTS

No book is the creation of the author alone. My thanks go to Phillip Earl, retired historian from the Nevada Historical Society, for his review of the manuscript and the Foreword, herein. Thanks also to good friends, all of whom have lived and/or traveled in Northwest Arizona in years past: Oyvind Frock, Bill and Mary Nork, and Barry and Norma Watson for their review of the manuscript.

Much appreciation goes to Paul Carson and Roseanne Rosenberg of the Mohave County Museum of History and Arts in Kingman, Arizona for all of their help in the research of this book.

And finally my wife, Ann, can take total credit for the layout and design of the book. With years of experience at the University of Nevada, Reno's in-house Printing Services, she made preparation of the book look easy.

AUTHOR'S NOTES

The idea for this book began after a spring 2003 visit to Oatman, Arizona and I discovered that there was little written in any one book about the history of Western Arizona. First ideas evolved, and slowly the areas of research expanded to cover Northwest Arizona. As with all manuscripts, many factors have to come together to produce a finished book.

History is a puzzle made up of many pieces. This book attempts to join the "pieces" together that make up the history of this area.

My education at the College of Mines and Geology and the Graduate School at the University of Arizona in the early 1960s, my career as an exploration geologist, plus my interest in amateur archaeology were all necessary ingredients in the formation of this manuscript, along with 25 years of writing and publishing four earlier books.

Research sources include the Museum of History and Arts in Kingman, Arizona; Getchell Library at the University of Nevada, Reno; miscellaneous books on the area; and utilizing internet information sources.

I found two books covering parts of Northwest Arizona that were outstanding and deserve recognition. The first is David Myrick's *The Santa Fe Route, Railroads of Arizona*, Vol. 4, Signature Press, Welton, California, 1998; and the second is by George H. Billinglsey, et al., *Quest for the Pillar of Gold*, Grand Canyon Association, Grand Canyon, Arizona, 1997.

Some historical facts are repeated in different chapters for continuity. And also, the following differences in the spelling of certain words or terms should be noted:

Within the text and maps, Hualapai (Walapai) refers to the Native American tribe; Wallapai is a mining district north of Kingman.

Pierce Ferry was established by Harrison Pearce.

The south rim of the Grand Canyon is capitalized in the manuscript after the Santa Fe Railroad arrived at the South Rim and the establishment of Grand Canyon Village (1901).

Mojave and Mohave are different spellings of the same word.

This book is intended for the residents of Northwest Arizona, those people interested in the area's history, and for those of us who pass this way and want to know more about the land and the people who shaped it.

ENJOY!!

CONTENTS

PART III — PLACES ON THE MAP

•

MAPS, CHARTS, ILLUSTRATIONS and PHOTOGRAPHS

The History of Mining in Northwest Arizona

The Railroad: The Atlantic & Pacific and Later, the Santa Fe

Man and the Grand Canyon

U.S. Route 66

The Civilian Conservation Corps in Northwest Arizona

PART III — PLACES ON THE MAP

Places on the Map

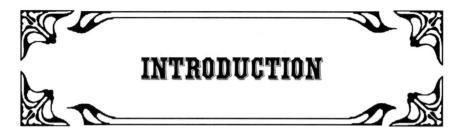

INTRODUCTION

This book is a journey through time — some 2.5 billion years — from the oldest rocks found in the Grand Canyon to the geologic evolution of the Colorado Plateau to man's first view of the Grand Canyon to the building of Interstate 40. It tells the story of Northwest Arizona: the land and the people who occupied it.

Northwest Arizona encompasses the area from the Colorado River on the west, the Arizona-Utah state line on the north, Interstate 40 on the south and U.S. 89 on the east. Its geopolitical development began with more than 3,000 years of Native American occupation.

After the explorations of Columbus in 1494, Spain and Portugal signed a treaty (Treaty of Tordesellas) that "divided all lands discovered in recent voyages of exploration and any others that may be discovered in the future between them."

Spain received all of the Western Hemisphere except Brazil. A little over a century later, Juan de Oñate founded the settlement of San Juan Pueblo in New Mexico, beginning the establishment of Spanish habitation throughout the southwest. With Mexican independence in 1821, these southwest Spanish territories became a province of Mexico.

Although the first American trappers and traders crossed northwest Arizona in the 1820s, it wasn't until 1848, after the Mexican War, that most of the region, including what is now Northwest Arizona, was ceded to the United States. It then became part of the New Mexico Territory, and attracted miners and ranchers along with the U.S. Army to provide protection. During the Civil War in 1864, the U.S. Congress created the Arizona Territory out of the western part of the New Mexico Territory and divided it into four political divisions including Mohave County.

With the discovery of gold and silver during the 1860s, thousands of miners, merchants and settlers added to the population. During the next 20 years U.S. Government surveys were conducted, wagon trails were constructed and the Atlantic & Pacific Railroad was completed.

Life was forever changed in the Southwest.

On February 14, 1912 Arizona was admitted to the Union as the 48th state.

Today, in addition to the state of Arizona, private parties and Indian Reservations, the U.S. Government owns significant amounts of land in northwest Arizona. These are administered by the Bureau of Land Management (BLM), the Forest Service and the National Park Service. The Santa Fe Railroad, as well, owns much property (in general, every other section for 20 miles on either side of their right-of-way).

Native Americans were the first to blaze foot trails across northwest Arizona. Much of their main Rio Grande-Pacific Ocean trail, used for trading, was followed by Beale's Wagon Road, the Atlantic & Pacific Railroad, The National Old Trails Highway, and U.S. Route 66. This opened northwest Arizona to cross-country travel, and millions of visitors now come from all over the world to enjoy the wonders of the Grand Canyon and the recreational diversity of this beautiful country.

Over the last 2,000 years a network of Native American trade-route trails was established linking the Pacific Ocean with the American Southwest. The main trail, the Rio Grande-Pacific Ocean trail, cut across northwest Arizona.

Looking northeast with the Black Mountains of Arizona in the background, the old Indian trail pictured above is just a few miles southwest of Needles, California and was part of the early trade route. Note the petroglyph on the boulder along the trail.

— Photo by Ann Ettinger.

PART I:
THE LAND

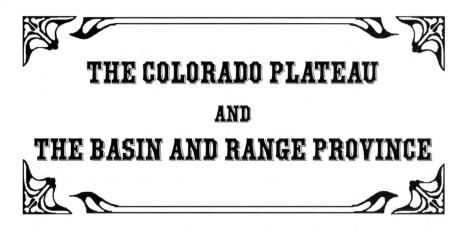

THE COLORADO PLATEAU
AND
THE BASIN AND RANGE PROVINCE

Stretching across most of northern Arizona and into southern Utah, south-western Colorado and northwestern New Mexico, the Colorado Plateau is a block of part of the earth's crust that has remained relatively intact for some 600 million years. Mountains and valleys of the Basin and Range Province mark the western and southern boundaries of the Colorado Plateau, and the Rocky Mountains form the northern and eastern boundaries.

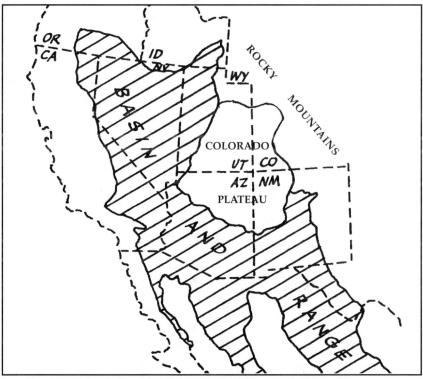

Location of the Colorado Plateau and the Basin and Range Province.
Modified from Fiero, 1986.

GEOLOGIC HISTORY OF THE COLORADO PLATEAU

Many studies have allowed a very complex geologic history of the Colorado Plateau to be pieced together. The oldest rocks underlying the Colorado Plateau are exposed in the Inner Gorge of the Grand Canyon. These pre-Cambrian rocks originated as a great thickness of marine sand, silt and mud which uplifted into a large mountain range about 2.5 billion years ago. As part of the mountain building process over hundreds of millions of years, the sedimentary rocks metamorphosed into schists and recrystallized into gneisses and granites, some of which intruded into overlying rocks as dikes and sills. The mountains eroded and the area subsided below sea level.

About 750 million years ago, younger pre-Cambrian limestones, shales and siltstones were deposited over the eroded roots of the earlier moun-

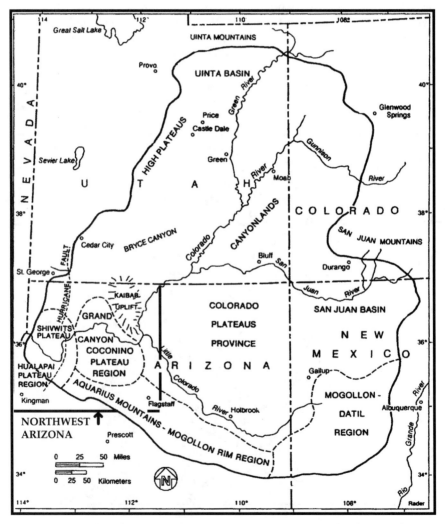

Detailed regions of the Colorado Plateau.   Modified from Graf, 1987.

tains. The area again uplifted into mountains, block-faulted and eroded over a period of many millions of years.

For most of the Paleozoic Era, from 545 to about 300 million years ago, there were two large continental crustal plates on the earth's surface: Laurasia (Laurussia), consisting of what would become North America, Europe and Asia; and Gondwanaland, consisting of what would become South America, Africa, Australia and Antarctica. During this time a large area which included the present-day Colorado Plateau submerged and a sedimentary sequence of more than 3,000 feet of limestones, shales and sandstones was deposited in fairly shallow advancing and retreating seas which bordered Laurasia, with two major periods of uplift above sea level and erosion. These marine sediments are now exposed in the Grand Canyon.

About 300 million years ago (Late Paleozoic) Laurasia and Gondwanaland collided when an area of what is now Africa crushed against what is now eastern United States and formed a megacontinent called Pangaea (Pangea). The movement of and collision between crustal plates is called plate tectonics and the driving forces are huge convection cells within the Earth's mantle.

Between 300 and 245 million years ago (Late Paleozoic) approximately 1,100 feet of sandy shale, siltstone, mudstone and limestone were deposited in a shallow near-shore environment, followed by uplift above sea level and an accumulation of about 1,000 feet of sand dunes, in a Saharan desert environment, into a large area that included the Colorado Plateau. A substantial period of uplift and erosion was followed by deposition of about 1,500 feet of sandstone, mudstone and shale deposited in a flood plain crossed by sluggish rivers.

About 225 million years ago (Middle Triassic) there is evidence that a great river, called the Chinle River by geologists, flowed westward over Pangaea from what is now the Texas Panhandle through New Mexico, Arizona, Utah and emptied into the ocean in what is now central Nevada. Little is known about the history of the Chinle River except that it finally changed course and flowed into the Gulf of Mexico. A large shallow basin then formed covering what is now southern Utah, northern Arizona, New Mexico and a part of western Texas. The basin slowly subsided and most of it remained just above sea level. Over millions of years some 2,000 feet of floodplain sediments consisting of mudstone, shale and siltstone were deposited in the subsiding basin in the area of the Colorado Plateau. This was followed by a long period of desert environment where about 2,000 feet of sand dunes interbedded with siltstones accumulated.

About 205 million years ago (Early Jurassic) Pangaea broke up and the present day continents formed. The shape of the land changed radically including upwarping and downwarping.

During the Early Jurassic (200 to 190 million years ago) southward wind-blown sands accumulated in the basin forming high dunes. These fossilized cross-bedded dunes (Navajo Sandstone), more than 1,000 feet in thickness, are seen today in Zion National Park and elsewhere in southern Utah as white and red-stained outcrops. By about 190 million years ago the area was uplifted and the land was exposed to severe erosion.

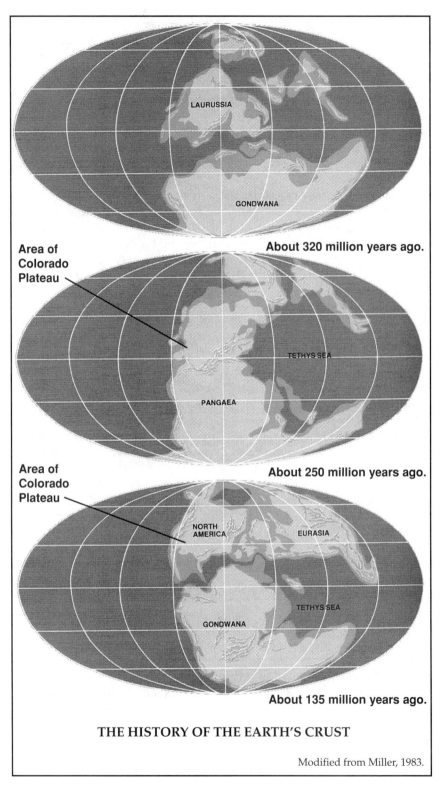

LAURUSSIA

GONDWANA

About 320 million years ago.

Area of
Colorado
Plateau

TETHYS SEA

PANGAEA

About 250 million years ago.

Area of
Colorado
Plateau

NORTH
AMERICA

EURASIA

TETHYS SEA

GONDWANA

About 135 million years ago.

THE HISTORY OF THE EARTH'S CRUST

Modified from Miller, 1983.

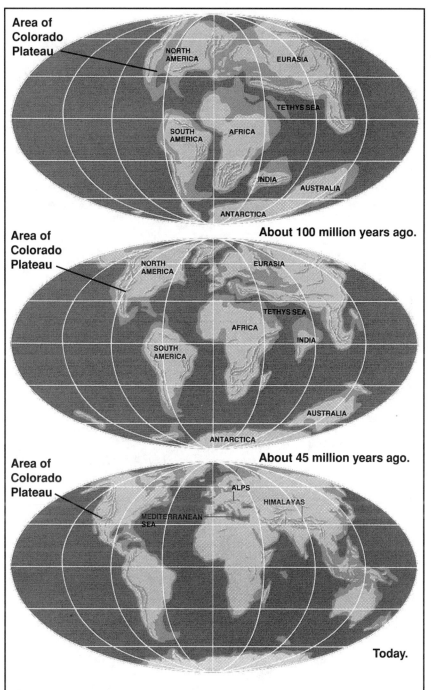

Area of Colorado Plateau

NORTH AMERICA

EURASIA

TETHYS SEA

SOUTH AMERICA

AFRICA

INDIA

AUSTRALIA

ANTARCTICA

About 100 million years ago.

Area of Colorado Plateau

NORTH AMERICA

EURASIA

TETHYS SEA

AFRICA

INDIA

SOUTH AMERICA

AUSTRALIA

ANTARCTICA

About 45 million years ago.

Area of Colorado Plateau

ALPS

HIMALAYAS

MEDITERRANEAN SEA

Today.

Geologists have been able to plot continental positions on the earth as far back as 550 million years. Examining fossil deposits and rock types reveals climatic conditions and ages of the rocks. Measuring residual magnetism at the time of rock formation indicates a continent's orientation and latitude relative to the North Magnetic Pole.

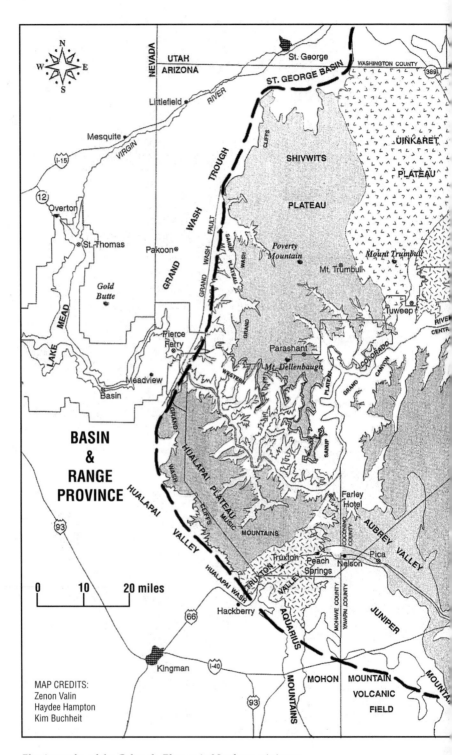

Physiography of the Colorado Plateau in Northwest Arizona.

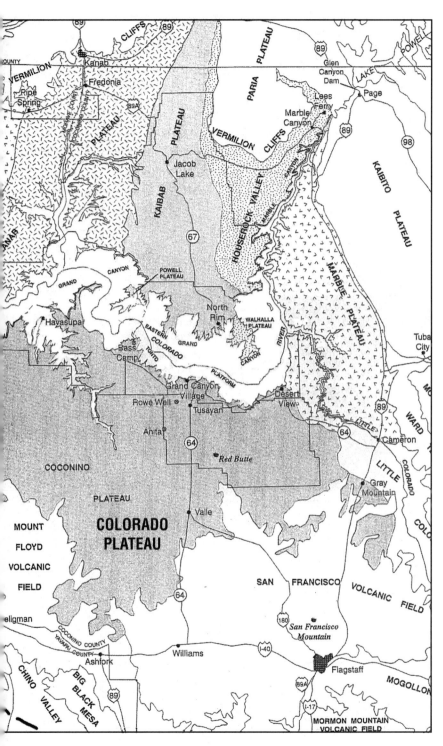

Modified from Billingsley, 1997.

GEOLOGIC TIME SCALE		ROCK DESCRIPTION	APPROX. THICKNESS (in feet)
ERA	PERIOD		
MESOZOIC 64-224 Million Years Ago	TRIASSIC	Brown conglomerate	25
		Light red to dark brown shale, siltstone and mudstone	400
		UNCONFORMITY	
PALEOZOIC 224-545 Million Years Ago	PERMIAN	Light gray fossiliferous limestone	300-400
		Fossiliferous limestone, siltstone, mudstone and sandstone	500
		Light tan sandstone formed from desert sand dunes	65-600
		Reddish shales, siltstones and mudstones formed in a swampy environment	100-200
	PENNSYL-VANIAN	Reddish sandstones, shales, siltstones and limestones formed as floodplain deposits	700-1600
		UNCONFORMITY	
	MISSISSIPPIAN	Gray limestone, stained red from overlying rocks	500-800
	DEVONIAN	Purple-pinkish limestone	100-450
		UNCONFORMITY	
	CAMBRIAN	Yellowish-gray limestone and siltstone	136-439
		Greenish-gray shale	270-450
		Brown sandstone	100-325
PRE-CAMBRIAN	PROTEROZOIC 545 Million to 2.5 Billion Years Ago	GREAT UNCONFORMITY Sandstones, limestones, shales and siltstones	12,000 ±
	ARCHEAN Over 2.5 Billion Years Ago	UNCONFORMITY Dark schists and gneisses intruded by granite	
		Modified from McKee, U.S.G.S. Professional Paper 1173, 1982.	

GEOLOGIC SECTION OF THE GRAND CANYON

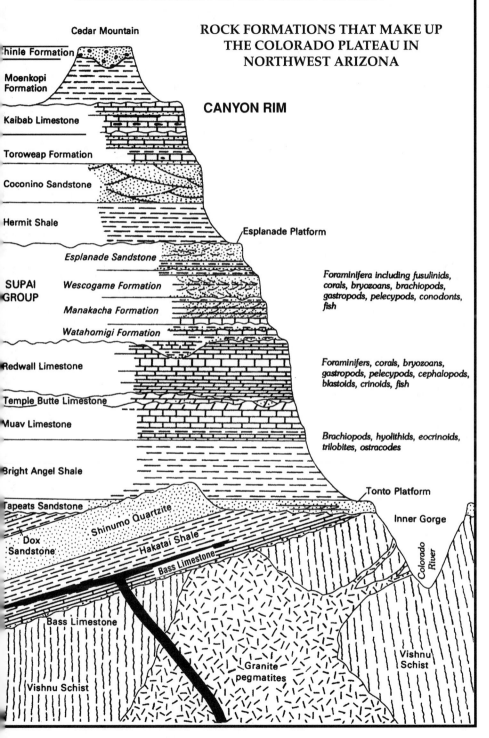

ROCK FORMATIONS THAT MAKE UP THE COLORADO PLATEAU IN NORTHWEST ARIZONA

Cedar Mountain

hinle Formation

Moenkopi Formation

CANYON RIM

Kaibab Limestone

Toroweap Formation

Coconino Sandstone

Hermit Shale

Esplanade Platform

Esplanade Sandstone

Foraminifera including fusulinids, corals, bryozoans, brachiopods, gastropods, pelecypods, conodonts, fish

SUPAI GROUP

Wescogame Formation

Manakacha Formation

Watahomigi Formation

Redwall Limestone

Foraminifers, corals, bryozoans, gastropods, pelecypods, cephalopods, blastoids, crinoids, fish

Temple Butte Limestone

Muav Limestone

Brachiopods, hyolithids, eocrinoids, trilobites, ostracodes

Bright Angel Shale

Tonto Platform

Tapeats Sandstone

Shinumo Quartzite

Inner Gorge

Dox Sandstone

Hakatai Shale

Colorado River

Bass Limestone

Bass Limestone

Granite pegmatites

Vishnu Schist

Vishnu Schist

Within the last 65 million years a vast area, including the Colorado Plateau, uplifted more than 10,000 feet above sea level by forces related to crustal plate movement. During this time of uplift of the Colorado Plateau, the processes of erosion stripped away almost all the Mesozoic rocks in northwest Arizona. What remains of the Mesozoic sequence is exposed in northeast Arizona and in southern Utah as part of the Grand Staircase consisting of, in ascending order, the Vermillion Cliffs, White Cliffs (Navajo Sandstone seen in Zion National Park), Grey Cliffs and Pink Cliffs (Tertiary Wasatch Formation seen in Bryce National Park).

Roughly 15 million years ago volcanic activity began when a "hot spot" within the upper mantle beneath what is now Seligman, Arizona fed molten rock through a weakness in the North American Crustal Plate resulting in the beginning of the Mount Floyd Volcanic Field (see Physiographic Map of the Colorado Plateau (pages 10 and 11). As the crustal plate slowly moved westward over the hot spot between 15 and 6 million years ago, volcanic activity continued between Seligman and Ash Fork.

Crustal movement over other "hot spots" is evidenced in the Hawaiian Islands, the Snake River Plains of southern Idaho, and into the Yellowstone area.

GEOLOGIC TIME SCALE		ROCK
ERA	PERIOD	DESCRIPTION
MESOZOIC	CRETACEOUS 140 million years ago	Found east and north of Zion. GRAY CLIFFS
		UNCONFORMITY
	JURASSIC (middle)	Sandstone and sandy limestone. Found in east and north part of Zion.
		——— UNCONFORMITY ———
	JURASSIC (early) 205 million years ago	White, cross-bedded dune sand. Cliff forming in Zion. WHITE CLIFFS
	TRIASSIC (late)	Interbedded sandstone, siltstone and mudstone deposited on a delta or flood plain. Sandstone equivalent of Wingate.
		Outcrops in Zion and Parunuweap canyons. UNCONFORMITY
		Shale and sandstone outcrops south and west of Zion. VERMILLION CLIFFS
		Outcrops in the southwest part of Zion. BELTED CLIFFS
		UNCONFORMITY
	TRIASSIC (early) 245 million years ago	Light red to dark brown shale, siltstone, and mudstone. Outcrops west and south of Zion. CHOCOLATE CLIFFS
		——— UNCONFORMITY ———
PALEOZOIC	PERMIAN	Outcrops in northern Arizona and the Grand Canyon.

WEST OF THE COLORADO PLATEAU:
THE BASIN AND RANGE PROVINCE

West of the Colorado Plateau, during Early Cenozoic (65 to 40 million years ago), all of the Paleozoic and Mesozoic rocks eroded off the pre-Cambrian basement. The Middle Cenozoic (45 to 25 million years ago) was a period of widespread volcanic activity. This was triggered when the North American crustal plate overrode the Pacific oceanic plate producing a subduction zone where the oceanic plate was forced deeper under the crust. This greatly heated the rocks which resulted in widespread volcanism west of today's Colorado Plateau. About 17 million years ago this land was a low-lying, almost featureless volcanic plain overlying a pre-Cambrian metamorphic basement complex. Then, between 15 and 7.5 million years ago,

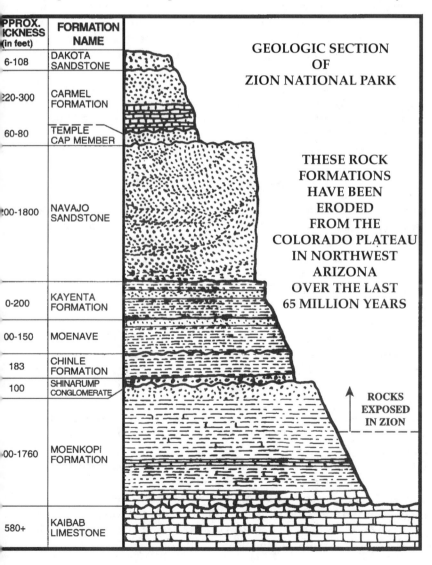

APPROX. THICKNESS (in feet)	FORMATION NAME	
6-108	DAKOTA SANDSTONE	**GEOLOGIC SECTION OF ZION NATIONAL PARK**
220-300	CARMEL FORMATION	
60-80	TEMPLE CAP MEMBER	
100-1800	NAVAJO SANDSTONE	**THESE ROCK FORMATIONS HAVE BEEN ERODED FROM THE COLORADO PLATEAU IN NORTHWEST ARIZONA OVER THE LAST 65 MILLION YEARS**
0-200	KAYENTA FORMATION	
100-150	MOENAVE	
183	CHINLE FORMATION	
100	SHINARUMP CONGLOMERATE	↑ **ROCKS EXPOSED IN ZION**
100-1760	MOENKOPI FORMATION	
580+	KAIBAB LIMESTONE	

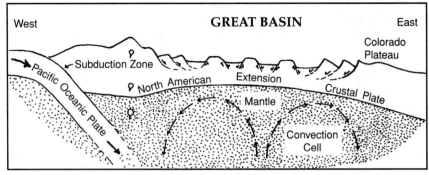

the crust began to be stretched and uplifted over an area 1,000 miles long and from 300 to 600 miles wide, probably the result of this part of the continent overriding the East Pacific Rise — a bulge in the oceanic floor. As the thinned crust arched upward, north-northeast, faulting occurred uplifting the mountains and downdropping the basins. This was the beginning of the Basin and Range Province and the processes of uplift, crustal thinning and faulting would last millions of years.

Ultimately, the area would be arched upward thousands of feet. During this time, erosion of the mountains has filled the valleys. This process of uplift and erosion is still going on today.

THE LAST SIX MILLION YEARS

Sometime about four million years ago the Ancestral Upper Colorado River was captured by a river to the west. A combination of uplift of the Colorado Plateau and the newly entrenched Colorado River of today allowed serious canyon cutting to begin. Evidence suggests that the lower 2,000 feet of the canyon were carved in the last 750,000 years.

Associated with uplift was scattered volcanic activity, some north-south faulting and monoclinal flexing, but in general the Colorado Plateau has remained essentially intact.

The same "hot spot" in the mantle, described on page 14, resulted in volcanic activity now seen as 9,233 foot Bill Williams Mountain, south of Williams. As the crustal plate continued to slowly move westward over the "hot spot," eruptions continued and produced more volcanoes progressively eastward.

Spanning the last six million years, more than 600 volcanoes have erupted in an area covering about 1,800 square miles between Williams and east of Flagstaff, north toward the Grand Canyon. Active between 2.8 million and 200,000 years ago, most of the volcanoes are basalt cinder cones less than 1,000 feet tall; a few are rhyolitic lava domes.

Sitgreaves Mountain is about 2.5 million years old. Mt. Kendrick (10,466 feet) is between and a few miles north of Sitgreaves Mountain and the San Francisco Peaks. The San Francisco Peaks are the remnants of a stratovolcano consisting of Mt. Humphrey (12,633 feet), Mt. Agassiz (12,356 feet),

GEOLOGIC TIME SCALE

MAJOR STRATIGRAPHIC AND TIME DIVISIONS IN USE BY THE U.S. GEOLOGICAL SURVEY

ERA	PERIOD	EPOCH	ESTIMATED AGES OF TIME BOUNDARIES (in millions of years)	DOMINANT LIFE FORMS
CENOZOIC	Quaternary	Holocene	0.01	Animals and plant of modern types.
		Pleistocene		
			3	
	Tertiary	Pliocene		
			7.5	
		Miocene		
			26	Age of mammals.
		Oligocene		
			37	
		Eocene		
			54	
		Paleocene		
			65	—— Mass extinction ——
MESOZOIC	Cretaceous	Upper (Late) Lower (Early)		
			140	
	Jurassic	Upper (Late) Middle (Middle) Lower (Early)		Age of reptiles, flowering plants, dinosaurs, first birds, mammals & modern fishes.
			205	
	Triassic	Upper (Late) Middle (Middle) Lower (Early)	220	—— Mass extinction —— Abundant coniferous trees.
			245	—— Mass extinction ——
PALEOZOIC	Permian	Upper (Late) Lower (Early)		Age of amphibians.
			250	
	Pennsylvanian	Upper (Late) Middle (Middle) Lower (Early)		Fern forests, abundant insects, first reptiles, large primitive trees.
			325	
	Mississippian	Upper (Late) Lower (Early)		
			355	—— Mass extinction ——
	Devonian	Upper (Late) Middle (Middle) Lower (Early)		Age of fishes, amphibians.
			415	
	Silurian	Upper (Late) Middle (Middle) Lower (Early)		Early plants and animals on land, shells dominant.
			440	—— Mass extinction ——
	Ordovician	Upper (Late) Middle (Middle) Lower (Early)		First vertebrates, first fish and insects.
			495	
	Cambrian	Upper (Late) Middle (Middle) Lower (Early)		Invertebrates dominant.
			545	
PRECAMBRIAN	Proterozoic			Primitive aquatic plants.
			2500	
	Archean			Bacteria and algae.

and Mt. Fremont (11,969 feet). They have been modified by Ice Age glaciation, erosion and possibly explosive eruptions. Northeast of Flagstaff is O'Leary Peak, a cluster of volcanic domes which pushed up between 240,000 and 170,000 years ago. South of O'Leary Peak is Sunset Crater, a cinder cone that formed during eruptions about 1,000 years ago and is now a national monument.

TODAY

Today we can see over 2.5 billion years of geologic history of the western Colorado Plateau exposed in the rocks of the Grand Canyon.

We see two time periods of mountain building and erosion in the lower Grand Canyon followed by millions of years of sedimentary deposition with periods of uplift and erosion.

Between 15 and 7.5 million years ago uplift of the Colorado Plateau began. About five million years ago the Lower Colorado River captured the Ancestral Colorado River, and the Colorado River of today began to cut the Grand Canyon creating one of the most scenic locations on Earth.

The condition of Meteor Crater near Winslow would suggest that there has been little erosion on the Colorado Plateau in northern Arizona since the meteor impact some 50,000 years ago.

The horizontal bedding of the marine sediments making up the Colorado Plateau can be seen from the highways of northern Arizona and from the rim of the Grand Canyon. The more resistant volcanic necks are also prominent features.

In the western part of Northwest Arizona, the exposed mountains of the Basin and Range Province consist of faulted blocks of pre-Cambrian metamorphic basement rocks. These are composed of granites, gneisses and schists, overlain or in fault contact with Tertiary volcanics. The environment is high to low Mojave Desert, depending on the elevation.

The range of climatic zones and plant environments on the Colorado Plateau and in the Grand Canyon area are as follows:

- Above 8,000': Spruce-Fir Community — 30 inches of precipitation yearly, subalpine forests of spruce and aspen on the North Rim and in the upper elevations of the San Francisco Volcanic Field;
- 6,500' - 8,000': Yellow Pine Community — sufficient rainfall and snow to support forests of Ponderosa pine, Douglas fir and aspen on both rims;
- 5,000' - 7,000': Piñon-Juniper Community — Pygmy forests of piñon and juniper trees on both rims of the canyon;
- 2,000' - 5,000': Desert-Scrub Community — high desert; and
- Below 2,000': Colorado River Community — low desert with less than 10 inches of rainfall annually along the Colorado River.

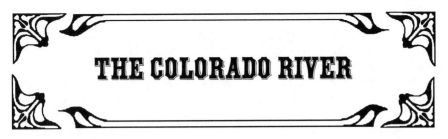

THE COLORADO RIVER

The Colorado River is the sixth longest river in North America, flowing across 1,360 miles in the United States and 90 miles in Mexico. Its history is an intriguing story.

Fray Francisco Tomas Garcés, a Franciscan missionary who viewed the river from the south rim of the Grand Canyon in 1776, called it the Rio Colorado, red-colored in Spanish, because of the brick-like hue of its silt-laden water.

The history of the Colorado River probably spans some 20 to 30 million years. Today the river flows from its headwaters in the mountains of Wyoming and into Colorado, southwestward through Colorado, Utah and Arizona, into Mexico and the Gulf of California. It drains an area of some

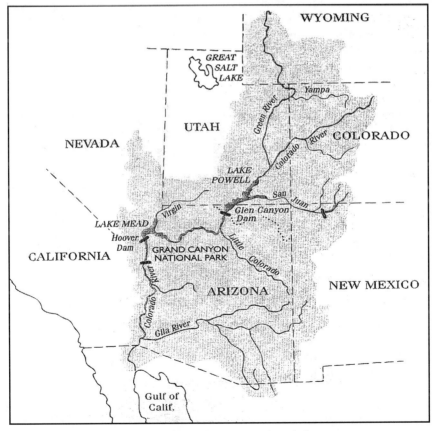

The Colorado River, its tributaries and its drainage area (stippled).
Modified from Price, 1999.

250,000 square miles, joined by the Green River in eastern Utah, the San Juan River in southeastern Utah, the Little Colorado River in northern Arizona, the Virgin River west of the Grand Canyon, and the Bill Williams and Gila rivers in southwestern Arizona.

From its headwaters the Colorado River falls some 10,000 feet in its course to the sea, flowing through 19 major canyons. The river's profile is V-shaped for the greater part of its course, a relative youth in the evolution of a river.

The Grand Canyon as we see it today was formed mainly from erosion by the Colorado River and its tributaries.

In the 277-mile length of the Grand Canyon the Colorado River, regarded as the roughest navigable river in the world, drops 2,200 feet and has more than 150 rapids.

Before construction of Hoover Dam (1936) and Glen Canyon Dam (1963), accurate estimates of the amount of silt and sand moved daily by the river for many years was 500,000 tons. In times of floods this figure could be 50 times higher. On September 13, 1927 a gauging station near the mouth of Bright Angel Creek in the canyon measured the Colorado River carrying 27.6 million tons of debris past that point. Add to that figure the amount of dissolved material and additional tonnage of pebbles, cobblestones and boulders that rolled along the bottom, the total material passing the gauging station that day may have been as much as 55 million tons.

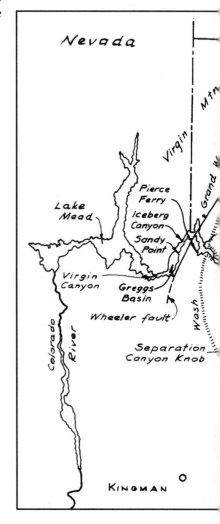

The overall history of the Colorado River includes its effect on the land it passes through and the people who came in contact with it. The geologic history spans more than 20 million years and includes the Ancestral Upper Colorado River, the much younger Lower Colorado River that captured the Upper Colorado River, the tectonics of the area, and the changes in climatic conditions which controlled the river's erosional power. The political history involves seven states, the U.S. Government and Mexico, and includes how to control the water flow, what to do with the water, and who gets how much. And finally, man's history with the river is concerned with crossing the river, traveling the river and recreation on the river.

THE ANCESTRAL UPPER COLORADO RIVER

Some 20 to 30 million years ago the lands of today's Utah, Arizona, Nevada, and southern California were relatively low-lying, probably less than 1,000 feet in elevation. Field studies over the last 30 years indicate that the Ancestral Colorado River originally flowed through Colorado, Utah, northeastern Arizona and into southern or central Nevada. This mature river crossed today's Kaibab Plateau near the present course of the Colorado River in the Grand Canyon, then turned northwestward toward the area of present-day St. George, Utah and then into present-day Nevada. Between 15 and 7.5 million years ago the Basin and Range Province formed, and the Ancestral Colorado River emptied into one or more of the basins. It is thought that this river was a dynamic entity changing through time. It evolved from segments of preexisting drainages with courses quite different from those seen today.

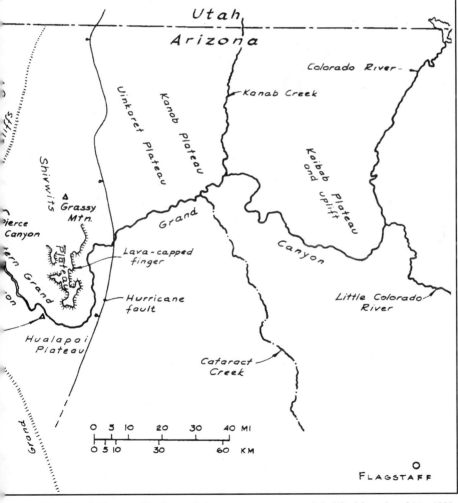

The Colorado River in Northwest Arizona.   Modified from Lucchitta, 1989.

EVOLUTION OF THE COLORADO RIVER: TWO THEORIES

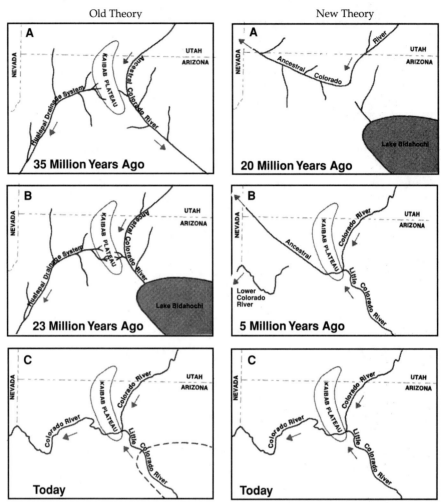

Old Theory   New Theory

Modified from Wallace, 1973.

Theories change with time. Thirty years ago, the evolution of the Colorado River showed the Ancestral Colorado River of 35 million years ago flowing into northern Arizona, blocked by the Kaibab Plateau, turning southeastward and flowing through New Mexico, then Texas or northern Mexico into the Gulf of Mexico. To the west of the Kaibab Plateau, the Hualapai Drainage System (Lower Colorado River) cut into the Kaibab Plateau.

About 23 million years ago, because of up†warping in eastern Arizona and western New Mexico, the Ancestral Colorado River formed large Lake Bidahochi, while the Hualapai eroded its way eastward.

About four million years ago the Hualapai eroded into the Kaibab Plateau and captured the Ancestral Colorado River.

The New Theory, based upon geologic field work tracing Ancestral Colorado River and tributary gravels northwestward across the Colorado Plateau, differs from the Old Theory in that the Ancestral Colorado River over many millions of years entered northern Arizona and then swung northwestward through northwest Arizona, possibly southwest Utah and into southern Nevada.

About five million years ago the Lower Colorado River eroded its way eastward, finally capturing the Ancestral Colorado River in the area of the Grand Canyon.

Streams that were probably tributaries to the Ancestral Colorado River flowed northward across what is now the western Grand Canyon as recently as 7.5 million years ago; i.e., the Grand Canyon and Lower Colorado River, as we know them, did not exist in the western Grand Canyon region as recently as 7.5 million years ago.

THE LOWER COLORADO RIVER
(Hualapai Drainage System)

Sometime after 7.5 million years ago the Lower Colorado River became established along its present course when the Gulf of California opened up and formed within a rift zone associated with the San Andreas Fault. The Lower Colorado River extended itself by headwater erosion and, sometime about four million years ago, the river captured the Ancestral Upper Colorado River.

A combination of nearly 5,000 feet uplift of the western Colorado Plateau over the last 7.5 million years and the entrenched Colorado River allowed serious canyon cutting. The river rapidly excavated the western Grand Canyon between five and four million years ago. Assisted by cycles of increased rain and snow associated with a number of glacial periods over several million years, the river greatly increased in size and erosional power. Over time, vast quantities of sediment carried by the Colorado River filled in a large part of the Salton Trough, the area of the Salton Sea in southern California.

In the last one to two million years there have been several periods of volcanic activity where basalt flows have dammed up the Colorado River in the western part of the Grand Canyon.

About a million years ago a lava flow dammed the Colorado River 1,400 feet high, forming a lake about 180 miles long. When the lava dams were breached, catastrophic flooding may have occurred downstream.

LOWER COLORADO RIVER STEAMERS

In 1852 the first Colorado River steamboat, "The Uncle Sam," traveled to Fort Yuma with supplies from a schooner anchored at Robinson's Landing near the mouth of the Colorado.

Three years later two steamers were supplying Fort Yuma and two years after that there was talk of opening up the river to steamer traffic all the way to Utah.

In 1858 Lt. Joseph Christmas Ives set out on a small stern wheeler, "The Explorer," to determine the navigability of the upper river. In 25 days Ives traveled more than 200 miles above Fort Yuma and concluded that the river was navigable to its junction with the Virgin River.

When the first immigrant party coming west on Beale's Wagon Road reached Beale's Crossing on the Colorado River in August of 1858, nine were killed in an Indian attack by a raiding party of Hualapai, including a few Mojaves. The War Department then ordered five companies of infantry to travel 200 miles overland, and 300 tons of provisions were sent up

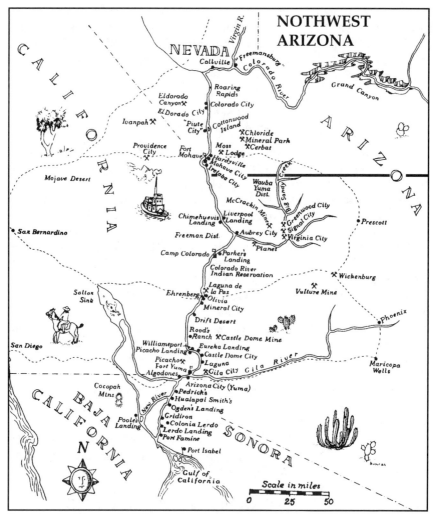

Steamboat landings on the Colorado River in the 1860s-1870s.
Modified from Lingenfelter.

the river by steamer to establish Fort Mohave. Two companies of the 6th Infantry and an artillery detachment under the command of Major Lewis A. Armistead built and manned the fort in mid-1859.

By the end of 1859 steam navigation of the Colorado was routine for a distance of 350 miles above its mouth.

The discovery of rich silver lodes in Eldorado Canyon in the spring of 1861, followed by the discovery of gold placers along the river, drew attention to the country bordering the river and triggered the Great Colorado River Rush of 1862. Thousands of miners, merchants and settlers poured into the country aiding the creation of the Arizona Territory in 1862.

This began a period of great activity in the mines and camps along the river — the golden age of the slim stern-wheelers that carried the supplies.

In 1864 William Harrison Hardy built a ferry crossing nine miles above Fort Mohave, and laid out the town of Hardyville along the Prescott and Mojave Toll Road.

The number of steamers working the river jumped from two in the spring of 1864 to five by the fall with the introduction of barges.

Occasionally steamers and barges became grounded on shallow bars but the steamboats were rarely aground for more than a week when a temporary rise in the river would free them.

For several years Hardyville was the principal shipping center for northern Arizona and for several years was the Mohave County seat (1870 to 1873). Only a few steamers a year ventured above Hardyville to supply Eldorado Canyon because Eldorado mines were mainly supplied by flatboats from Mormon settlements upriver. These flatboats were broken up and sold as firewood upon arrival.

In 1877 the Southern Pacific Railroad reached Yuma, taking away the steamer business on the Lower Colorado, but steamers continued to haul supplies north of Yuma. Steamer trade declined with the arrival of the Atlantic & Pacific Railroad at Needles in 1883.

In the late 1880s declining silver prices forced many mines in the area to close, further hurting the steamer business.

In the early 1890s new gold ore discoveries were made along the Colorado River above Hardyville and the river trade prospered again with the renewed mining activity.

In 1906 a fleet of gasoline boats replaced the side-wheel and stern-wheel steamers.

The era of steam navigation on the river ended with completion of the Laguna Dam, which closed the river 14 miles above Yuma in 1909. The last steamboat on the river was reported "lost on the river" on October 3, 1916.

UPPER COLORADO RIVER STEAMERS

The vast canyon country of the Upper Colorado was settled much more slowly than the Lower Colorado. In 1868 the Union Pacific established a station at Green River, Wyoming and in 1869 Major Powell began his explorations of the canyon country from there.

The first steamboats in the canyon country were in 1890. They operated until 1905 when they were replaced by boats with gasoline engines.

TAMING THE COLORADO RIVER

Because of its vast drainage area the Colorado River has, since its beginning, periodically flooded. Flooding really became a problem when steamships sailed up the Lower Colorado in the 1860s and later when the Atlantic & Pacific and Santa Fe railroads began building bridges across the river.

The answer was construction of dams along the Colorado River for flood control, water storage, and hydroelectric power generation. In 1903 E.T. Perkins conducted the first dam survey along the river, recommending sites at Boulder, Black Canyon, Bullhead and Parker.

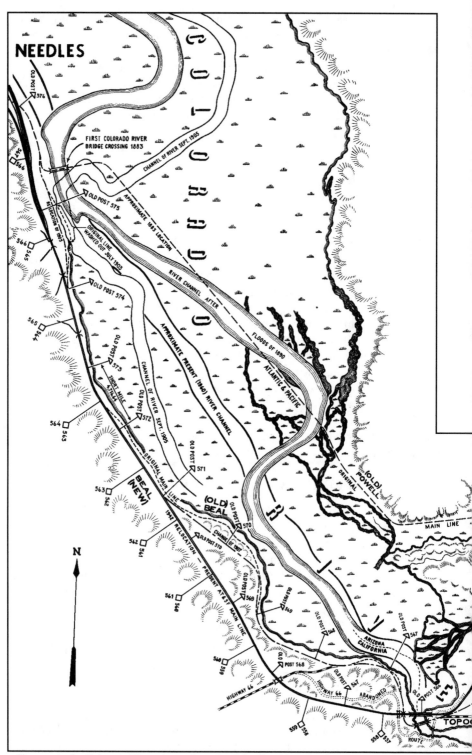

Changes in the Colorado River channel south of Needles. Note the changes in the Atlantic and Pacific and later Santa Fe railroad lines due to the changes in the river channel.

In the early 1900s there was little environmental concern and environmental impact statements didn't exist. Therefore, building dams was politically easy.

In the early 1900s several large floods devastated land along the Lower Colorado. A major flood occurred in 1905 when the river broke through irrigation headgates and flowed north into the Salton Basin in southern California to form the Salton Sea, a lake about 240 feet below sea level and about 400 square miles in size. In 1907 the headgate breech was sealed at the cost of millions of dollars and the river again followed its original course into the Gulf of California.

These natural disasters, along with questions of how to divide up the water of the Colorado among the seven states within the Colorado River Basin, began a political process which would determine the destiny of the river.

After the flood of 1905 which created the Salton Sea, and for a number of years thereafter, the construction of extensive levees on the Lower Colorado was tried, but was not an answer to flood or irrigation control. After numerous government reports and several years of engineering investigations, Congress authorized the Boulder Canyon Project of 1928.

Construction of the Boulder Canyon Project dam was authorized in 1928 to control floods, improve navigation, provide for storage and delivery of water and generate electrical energy to help the project be self-supporting. The site selected was in the Black Canyon of the Colorado River about 25 miles southeast of Las Vegas, Nevada.

The Bureau of Reclamation designed the dam and supervised its construction under contract to a joint venture of six of the largest dam builders

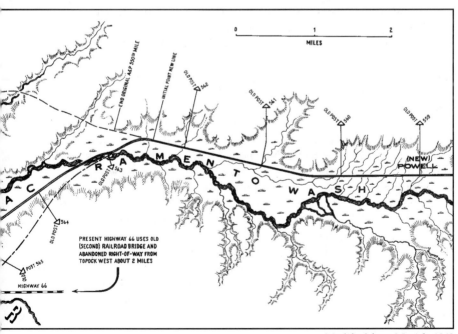

Modified from Myrick, 1963.

in the country. The original construction campsite housed an average of 3,500 workers at what is now Boulder City, Nevada.

First called Boulder Canyon Dam, construction began in 1932 and was completed in 1935. The dam is 726 feet high and 1,244 feet long, with a base of 660 feet and contains more than 4.4 million cubic yards (six million tons) of concrete. The dam cost $120 million and the entire project about $385 million. In 1931, the dam was unofficially called Hoover Dam to honor President Herbert Hoover. The name was changed to Boulder Dam in 1933. In 1947 Congress officially named it Hoover Dam.

The highest dam in the United States created Lake Mead, 115 miles long and 589 feet deep with a storage capacity of 10 trillion gallons (32.3 million acre-feet) of water. It took six years to fill Lake Mead.

To control flooding, water distribution and to generate hydroelectric power along the Lower Colorado, three new dams would be constructed downstream from Hoover Dam in the 1930s and 1940s as part of the Lower Colorado Dams Project.

Construction of Parker Dam, about 145 miles south of Hoover Dam, began in 1934 and completed in 1938. The concrete arch dam, 320 feet high and 856 feet long, formed Lake Havasu and was built to supply water and electric power to southern California.

Davis Dam is 67 miles south of Hoover Dam and two miles north of Laughlin, Nevada. The 200 feet high, 1,600 feet long earth and boulder-fill dam was constructed between 1942 and 1953 at a cost of $67 million as part of the Parker-Davis Project.

Imperial Dam, a 3,475 foot long concrete dam, was constructed 28 miles northeast of Yuma, Arizona (181 feet above sea level) between 1936 and 1938, to divert water into the All American and Coachella Canals to southern California. There is a large de-silting facility in proximity to the dam.

In 1956 Congress passed the Colorado River Storage Act, providing for construction of four major water storage and hydroelectric power generating dams on the Upper Colorado River and its tributaries. Construction under the direction of the U.S. Bureau of Reclamation began on the Glen Canyon Dam in 1956 and was completed in 1964. Lake Powell took 17 years to fill to its full capacity (3,700 feet above sea level) with water derived from snowmelt from the western slopes of the Rocky Mountains.

Between 1958 and 1963 the Navajo Dam on the San Juan River and the Flaming Gorge Dam on the Green River, both Colorado River tributaries, were constructed as part of the Colorado River Storage Act.

In 1963 plans were made to develop recreation areas and wildlife preserves along both banks of the Lower Colorado River between Hoover Dam and the Mexican border.

These dams were all built before environmental impact statements were required by law (1969), and it has been learned that there is a trade-off of advantages versus interests, one of which is that dams change the natural flow of the river affecting the plant and animal environment.

Engineers or surveyors along the Colorado River at Diamond Creek Canyon west of the Grand Canyon, circa 1915.
Courtesy of the Mohave Museum of History and Arts, Kingman, Arizona.

THE COLORADO RIVER COMPACT

The question of water distribution among the seven Colorado River Basin states persisted and finally delegates from Colorado, Utah, Wyoming, New Mexico (Upper Basin States), Arizona, Nevada and California (Lower Basin States) met on November 9, 1922 in New Mexico to discuss and negotiate how Colorado River water would be apportioned.

Known as the Colorado River Compact, water was "apportioned from the Colorado River in perpetuity to the Upper Basin and Lower Basin" in the amount of 7.5 million acre-feet of water per year to each basin. It was agreed that the arbitrary boundary of the Upper and Lower Basins would be Lee's Ferry near the Utah-Arizona state line and that the states within each basin would negotiate each state's allocation. It wouldn't be until 1948 that the Upper Basin States agreed on their water allocation. The Lower Basin States water allocation was finally decided by a U.S. Supreme Court decision in 1963.

The Colorado River Compact did not allow any water for Mexico, so it was modified in 1944 in the Mexican Water Treaty to guarantee 0.73 million acre-feet of water per year to Mexico. This meant that the Upper Basin was required to deliver 8.23 million acre-feet of water each year to the Lower Basin states and Mexico. There have been years when rain and snow precipitation has been below average in the Upper Basin drainage area, and the Upper Basin states have had problems delivering the agreed amount of water. This has led to disputes between states and between the United States and Mexico.

Indian reservation rights to Colorado River water were not considered in the Colorado River Compact in 1922. A 1908 U.S. Supreme Court decision recognized Indian water rights and the state in which a reservation is located must fulfill those tribal water rights. Indian reservations in the Upper Basin have been granted about one million acre-feet of water through federal legislation and in litigation. A part of the 1963 U.S. Supreme Court decision that determined the Lower Basin water allocation also quantified the water rights of the five Indian reservations along the Lower Colorado River.

In subsequent years Congress has legislated other laws affecting the Colorado River. These include the Endangered Species Act of 1973 (which protects endangered fish in the river), the Colorado River Basin Salinity Control Act of 1974 (which deals with salinity and water quality), and the Grand Canyon Protection Act of 1992 (which recognized the recreational value of the Colorado River to the Grand Canyon National Park).

PART II:
THE PEOPLE

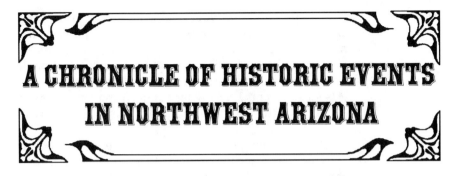

A CHRONICLE OF HISTORIC EVENTS IN NORTHWEST ARIZONA

Early Man   Native Americans of the Desert Archaic culture were likely the first occupants in northwest Arizona about 4,000 years ago.

Two to three thousand years ago peoples of the Patayan culture entered northwest Arizona. They were the ancestors of today's Mojave, Havasupai, Hualapai and Yavapai tribes. About 400 A.D. the Anasazi expanded into the Grand Canyon area and remained until about 1300 A.D. when they abandoned their land.

1540   Francisco Coronado, searching for the Seven Lost Cities of Cibola, made the initial European contact in the region that is now Texas and New Mexico, establishing Mexico's claim to the area. Other Spanish explorers followed, seeking precious metals.

1598   Juan de Oñate, a mine owner, led an inland expedition from what is now Mexico to establish a permanent land base that would protect New Spain's northern frontier. On April 30, at a site near what is now El Paso, Oñate proclaimed Spain's sovereignty over all the "lands in the kingdom and province of New Mexico," a vaguely defined wilderness stretching all the way from present-day Texas to California.

With his expedition, Oñate brought 7,000 Spanish cattle, sheep, goats, burros, and a herd of the first modern-day horses into what would become the United States.

Oñate crossed what is now Arizona in 1604.

The Spaniards had gold fever and over the years they prospected much of the southwest, finding little gold.

1775-1776   Fray Francisco Garcés, a Franciscan missionary, traveled through the Oatman/Kingman area on his journey to bring Christianity to the Yavapai Indians of Arizona.

Padres Silvestre Escalante and Francisco Domiguez, with a small band of soldiers and guides, explored the country north of the Colorado River. Their attempt to cross the Colorado River near the mouth of the Paria River failed. The expedition later crossed the river to the east in Utah.

1820s	A few trappers and fur traders crossed the area. In 1826 and 1827 Jedediah Smith twice crossed the western part of what is now Arizona from a point where the Virgin River meets the Colorado River, overland along the Colorado to a point north of Needles, and then went west to Los Angeles.
	Also in 1826 and 1827 Ewing Young, a trapper, followed the Colorado River on horseback from Yuma all the way to its headwaters in Colorado.
	In 1829, following Jedediah Smith's route to the junction of the Virgin River and the Colorado River, Peter Skene Ogden of the Hudson Bay Company explored northwest Arizona.
1844	Fremont's second expedition traversed what is now the northwest corner of Arizona into Utah.
1848	The Treaty of Guadalupe Hidalgo brought the portion of what would become Arizona lying north of the Gila River, including the Mohave County area, into the United States. Miners and settlers then started to move into the territory.
1850	The New Mexico and Utah territories were established.
1852	Captain Lorenzo Sitgreaves, one of the first U.S. Topographical Engineers, and party explored the Oatman/Kingman area to survey a possible wagon road along the 35th Parallel across Arizona following centuries old Indian trails.
1853	Referred to as the Gadsden Purchase, the United States purchased land south of the Gila River from Mexico which would become part of the New Mexico Territory.
1853-1854	Lt. Amiel W. Whipple surveyed and mapped a railroad route which followed the 35th Parallel from Ft. Smith, Arkansas to the Pacific Ocean.
1854	Francois Xavier Aubrey was the first man to take a wagon from San Jose, California along the 35th Parallel across northwest Arizona and then to Santa Fe, New Mexico, following the Sitgreaves and Whipple survey routes.
1857-1858	Lt. Edward Beale, commissioned by the government to lay out a wagon road that could support the passage of heavy wagons along the 35th Parallel, retraced much of Whipple's survey. Camels, imported from the Middle East as pack animals, were used as the "camel experiment." In addition to pack mules and horses, the camels were found to be well suited to surviving in the western desert environment.
	Beale's road building crews located springs in the Kingman area that would eventually bear his name and become one of the first water sources for the new town of Kingman. The Beale Road, which stretched from Fort Defiance, New Mexico

to the Colorado River, became a popular route for prospectors seeking gold, silver, copper and turquoise.

Beale crossed a pass in the Black Mountains a few miles east of present-day Oatman in October 1857. In January 1858 he crossed the pass again and named it John Howell's Pass.

Lt. Joseph Christmas Ives and his men steamed 200 miles up the Colorado River from Fort Yuma in a small stern wheeler, the U.S.S. Explorer. The vessel smashed on rocks, after which Ives and his men continued overland.

Five months later, Ives crossed John Howell's Pass and renamed it Sitgreaves Pass after the leader of the first expedition into the area.

The U.S. Explorer on the Colorado River in 1858.
Courtesy of the Mohave Museum of History and Arts, Kingman, Arizona.

1859	Prospectors began trickling into the area. Fort Mohave was established by Col. William Hoffman, 6th U.S. Infantry, as protection against the Indians. Major Lewis Armistead was placed in charge when Col. Hoffman was ordered elsewhere by his superiors.
1861	Fort Mohave was abandoned at the beginning of the Civil War.
1862	Jacob Hamblin crossed the Colorado River west of the Grand Canyon at the site that would later become Pierce Ferry.
1863	Congress created the Arizona Territory which was carved out of the western part of the New Mexico Territory. Fort Mohave was re-garrisoned by two companies of Fourth California Volunteers.
1864	Mohave County was established with the first county seat at Mohave City.

Men from Ft. Mohave, along with other prospectors, explored for gold in the Black Mountains. John Moss discovered a rich deposit about 5.5 miles northwest of present-day Oatman. When silver was found in the Cerbat Mountains most of the diggings in the Black Mountains were abandoned.

Captain W.H. Hardy established a trading post and a ferry landing, which became known as Hardy's Landing, and later Hardyville. The incorporation of the Prescott and Mojave Toll Road was approved by the Legislative Assembly. A station was established at Beale's Springs and the toll road continued westward over Union Pass to Hardyville on the Colorado River.

Following a Brigham Young directive to settle along the Little Colorado River in Arizona, in March 1864 Mormon frontiersman Jacob Hamblin and 15 men built a raft and made the first successful crossing at a point on the Colorado River that would become Lee's Ferry. For the next several decades thousands of Mormon settlers would homestead in northern Arizona.

Between 1864 and 1883 horses, wagons, stage lines and Colorado River steamers were the main modes of transportation.

1865   The Prescott Toll Road paid the Hualapai tribe $150 in merchandise for rights to cross their land.

1867   The federal Townsite Act of 1867 amended the Townsite Act of 1844, whereby the United States government permitted public lands to be settled upon and occupied as townsites.

1869   Major John Wesley Powell made his first expedition down the Colorado River through the Grand Canyon.

1870s   More mines were discovered in the Cerbat Mountains some 20 miles north of Kingman, and the towns of Cerbat, Chloride and Mineral Park appeared.

Cattle were driven into the area for the rich grazing land and ranchers raised enormous herds. They sold their beef to agents for Indian reservations and to residents in the growing mining camps and settlements.

1871   Lt. George Wheeler of the U.S. Corps of Engineers surveyed the Kingman area on his mapping expedition of northern Arizona.

Camp Beale's Springs was commissioned and served as a temporary reservation for the Hualapai Indians. The Camp was deactivated in 1874.

1872   The federal Mining Law of 1872 set standards and guidelines for mining districts and mining claims.

1873   Lee's Ferry operations across the Colorado River began and lasted until 1928. Crossing fees for Mormon travelers were $2.00 per wagon, $1.00 per horse and rider, and $.04 per head of stock; non-Mormons paid about 50% more.

1874	There was weekly mail from Prescott to Camp Beale Springs, Cerbat, Mineral Park, Chloride and Hardyville.
1877	Mohave County had two schools with 46 pupils.
1880	Lewis Kingman began surveying a railroad route between Albuquerque, New Mexico and Needles, California, which paralleled Beale's Road for much of its length.
	Many of the railroad camps and sidings grew into small communities and a few — Flagstaff, Kingman and Needles — grew into small cities.
	Fort Mohave Indian Reservation was established on September 19th.
1882	Kingman (first known as Middleton) was established as a siding on the new Atlantic & Pacific Railroad (later to become the Santa Fe Railroad). A rooming house, stores, and other buildings were built. This would be a shipping center for ranches, miners and Hualapai Indians.
1883	The railroad reached the Colorado River which was bridged, linking the Midwest with California. With the completion of the Atlantic & Pacific Railroad in 1883, large ranches and logging companies could easily ship cattle and lumber to distant markets.
	Prospectors arrived at the Grand Canyon and quickly realized that there was more money to be made with tourists than mining.
1887	Kingman became the county seat of Mohave County.
1890	Mohave County's population was about 1,500 with 300 at Kingman. Cerbat had a five-stamp gold mill and a five-stamp silver mill. Gold Basin had a 10-stamp mill and Signal a 10- and 20-stamp mill.
	The cattle industry in Mohave County was thriving. About 60,000 head of cattle and about 5,000 goats ranged in the valleys, mountains, and on the mesas.
	Two thousand acres were under cultivation (alfalfa, barley and vegetables) in the valley of the Big Sandy.
1893	Land in Mohave County was selling for $6 per acre.
1895-96	Northwest Arizona suffered a debilitating drought. It is estimated that 40 to 60% of the range cattle perished for lack of water and forage. Thousands of cattle and sheep were shipped out of the territory on the A&P Railroad.
1897	The Santa Fe Railroad took over the Atlantic & Pacific Railroad.
	A week-long storm washed out roads and railbeds.

1899	New ore deposits were found in the Oatman area by Joe Jerez.
1900	Gold was discovered at Gold Road in May. Kingman, with a population of 500, became a center for mining activities in Mohave County.
1901	The Santa Fe Railroad built a 65-mile spur line from Williams to the South Rim of the Grand Canyon.
1902	The Oatman-Gold Road mining area opened up.
1903	The first dam survey was made along the Colorado River by E.T. Perkins, recommending sites at Boulder, Black Canyon, Bullhead and Parker.
1905	Grand Canyon Village had its beginning on the South Rim when the Fred Harvey Company built several lodges, including the El Tovar Hotel, the Hopi House and gift shops.
1912	Arizona became the 48th state on February 14th.
1914	The National Old Trails Highway was completed essentially following Beale's Wagon Road, winding its way through Kingman over Sitgreaves Pass through Oatman and on to Needles. With it came a small but ever-increasing amount of automobile traffic.
1916	An automobile bridge was completed across the Colorado River at Topock.
1919	The Grand Canyon was designated a National Park. Kingman boasted seven garages, three meat markets, two drug stores, two churches, a Western Union, two lumber yards, a picture show, numerous hotels and saloons, and a Yucca Fiber Factory which made rope from the Yucca plant.
1921-26	Six years of drought resulted in thousands of cattle being shipped by rail to other areas.
1926	A new federal highway, essentially following and improving the National Old Trails Highway, was designated through Kingman; U.S. Route 66 began its life as the "Main Street of America."
1928	The Boulder Canyon Dam Project was authorized by Congress.
1929	The Kingman Airport was dedicated on June 25. The Great Depression began and lasted until the onset of World War II.
1931	The road from Kingman to the Boulder Dam site was completed (now U.S. 93).
1933-42	About 12 Civilian Conservation Corps (CCC) camps were

Ford Tri-motor at the Kingman Airport dedication, June 25, 1929.
Courtesy of the Mohave Museum of History and Arts, Kingman, Arizona.

established and operated in northwest Arizona. They worked on projects in the Grand Canyon, on U.S. Route 66 in the Kingman and Flagstaff areas, in parks and on the open range. These paramilitary camps each housed 200 to 250 unskilled and unemployed young men between the ages of 18 and 25 who were provided food, shelter, clothing and employment with the opportunity to learn a trade.

Projects included construction of cattle fences, stock tanks, road building and improvements. They treated 50,000 acres for rats, developed springs, improved parks (trails, picnic areas and covers), built dams across washes, bridges and cabins and conducted archaeological excavations, including a turquoise mine, ancient house, camp, rock shelters and a cave.

1935   The federal Taylor Grazing Act regulated cattle and sheep on public lands.

1935-36   Hoover Dam was dedicated by Franklin D. Roosevelt on September 30, 1935 and the powerplant structures were completed in 1936. Lake Mead became the largest man-made lake in the United States and a popular recreation area.

1936   Parker Dam was completed forming Lake Havasu.

1938   Route 66 was completely paved.

1941   Davis Dam construction was authorized. Bullhead City was begun by Utah Construction Company (prime contractors for Davis Dam) as a construction city.

1941-1945   The U.S. entered World War II when the Japanese attacked Pearl Harbor. The war lasted until 1945.

The impact of the war on Northwest Arizona included the construction of Kingman Army Air Force base, called Kingman Arizona Airfield, and the Navajo Ordinance Depot at Bellemont. Santa Fe trains carried men and equipment to the West Coast, and Route 66 was used for military convoys going west.

1942   Franklin Delano Roosevelt signed Executive Order L-208 which stopped all mining activities which were nonessential to the war effort.

In 1942 the Desert Training Center, California-Arizona Maneuver Area, was created by the War Department to train troops for battle areas in North Africa and the Mediterranean. Initially under the command of General George C. Patton, this was the largest military training ground in the history of military maneuvers. The Arizona part of the training center, called "Area C," covered an area west of U.S. 93 to Kingman and north of present-day I-40. The project closed in 1944.

Much of the same area was reused in May 1964. It was called "Operation Desert Strike," a joint-branch combat simulation during the height of the Cold War, where troops from "Nezona" invaded "Calonia."

1952   A new section of Route 66 was opened between Kingman and Topock, bypassing Oatman.

1953   Davis Dam was completed.

1956   At Yucca, 24 miles south of Kingman along Route 66, Ford Motor Company set up a vehicle proving ground on 4,000 acres of an abandoned Army air base. This was used to test cars and trucks for performance and durability in a hot, dry climate.

1960   Duval Mining Corporation began a large open-pit copper-molybdenum operation at Mineral Park north of Kingman.

The Santa Fe Railroad relocated 44 miles of their main line between Williams and Crookton.

1963   Glen Canyon Dam was completed and Lake Powell became the second largest man-made lake in the United States.

1968   Santa Fe Railroad passenger service to Grand Canyon Village ended.

1970-1977   Grand Canyon was mapped in detail by the U.S. Geological Survey.

1980   U.S. Interstate 40 opened through Northwest Arizona.

1989   The Grand Canyon Railway from Williams to the South Rim reinstated passenger service.

NATIVE AMERICANS IN NORTHWEST ARIZONA

Today five Native American tribes occupy reservations in northwest Arizona: the Mojave (Ahamakav, "People Who Live Along the River"), the Havasupai, the Hualapai (Walapai), the Kaibab-Paiute and the Navajo. The earliest Native American occupants of northwest Arizona belonged to the Desert Archaic Culture, followed by the Patayan and Anasazi cultures.

PALEO-INDIAN AND EARLY ARCHAIC INDIAN

Split-twig deer talisman from the Grand Canyon (circa 2500 B.C.). From Waldman, 1985.

Even though no sites have been found, it is probable that small numbers of Early Archaic hunter-gatherers crossed northwest Arizona from time to time as far back as 7,000 years ago.

Archaeological evidence suggests that Native Americans of the Desert Archaic Culture were the first to descend into the depths of the Grand Canyon some 4,000 years ago. They were nomadic hunter-gatherers living in small family groups. They left evidence of their presence in caves in the Redwall Limestone, including small split-twig figurines. For these early peoples the Colorado River was a formidable boundary.

THE PATAYAN

The Patayan were ancestors of the Mojave, Havasupai, Hualapai (Walapai) and Yavapai Tribes.

Peoples of the Yuman-speaking Patayan culture first entered northwest Arizona some two to three thousand years ago. Residing along the Lower Colorado River were the River Yuman, later to become the Mojave Tribe. South of the Grand Canyon in northwest Arizona were the upland Yuman-speaking Pai. Originally made up of 13 regional bands, they were sparsley populated families of hunter-gatherers just beginning to farm.

The early Pai lived in above-ground domed shelters, thatched with grass. Their crops included a variety of grasses in the river and stream floodplains which were planted in late spring. They tended their crops through most of the summer. Their gathering activities included mesquite, which may have been about 50% of their diet, and also seeds, roots, berries and nuts.

Later the Pai would winter in the mountain foothills in pithouses which served as the base for hunting and gathering activities. Hunting included rabbits, rodents, deer, antelope, mountain sheep, and a variety of birds. Fish were a part of their diet where fishing holes existed. Their craft items included mats, baskets, blankets and pottery. They were active traders and maintained an elaborate network of trails across their territory.

The Pai first appeared in the plateau region south of the Grand Canyon about A.D. 600.

Pai sites dated between A.D. 655 and A.D. 1300 have been found in the Cerbat Mountains north of Kingman. Population, small at first, tripled by A.D. 800 and doubled again by A.D. 900. By A.D. 1050 farmland on the canyon floors began to be cultivated on a permanent basis during the spring and summer.

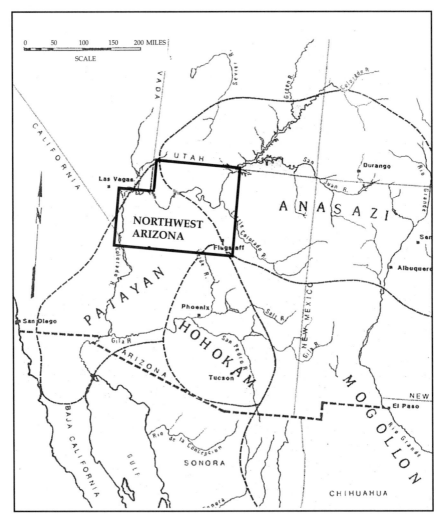

Area of Patayan (Pai) and Anasazi influence about A.D. 1000 in Northwest Arizona. Modified from Cordell, 1984.

Between A.D. 1050 to A.D. 1200 the Pai that occupied the Coconino Plateau relocated to Cataract Creek Canyon for what appears to be defensive purposes. Like the Anasazi in northeastern Arizona, they built cliff dwellings in which they resided until about A.D. 1300, when they returned to the Coconino Plateau, including the Grand Canyon area, for their hunting and gathering.

It is thought that the Pai originally migrated eastward from southern California. They were the ancestors of today's tribes found in northwest Arizona, sub-grouped under the names of Mojave, Hualasupai, Havasupai and Yavapai. The Havasupai and Hualasupai were distinguished by name as early as 1776, and were perhaps distinct much earlier. Until the mid-1880s, both the Havasupai and the Hualasupai carried on hostile relations with the Yavapai, and there were frequent raids during the harvest season. There was an agreed upon end to hostilities and there has been peaceful trade relations since then. Today these tribes are politically separate groups, but are ethnically one people of the same basic culture.

The Pai lifestyle before the mid-1800s was one of farming, hunting and gathering. Planting began in mid-April. They harvested corn beginning in June and continued until fall, at which time all crops and many kinds of wild plant foods had been picked and processed for storage. Drying was the usual method used in preserving food for winter use. By the middle of October families began moving back to their semi-permanent camps which

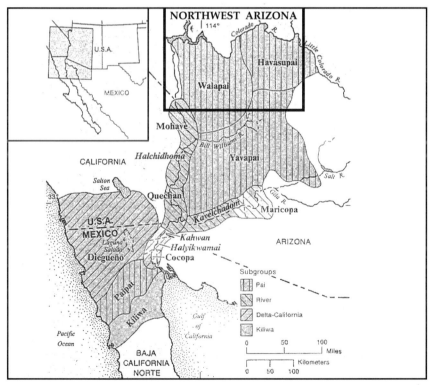

Approximate distribution of Pai Tribes at time of European contact.
Modified from Kendall, 1983.

were scattered over the plateau. They spent the winter hunting deer, antelope, rabbit and gathering piñon nuts, mescal and other wild plant foods.

Camps were moved when necessary to obtain new resources. With the coming of spring, the planting cycle began again and they migrated back to their summer homes near the river and stream floodplains on the canyon floor.

Trade was another important economic activity. Buckskins, foodstuffs and basketry were traded for cotton goods, horses, pottery, jewelry and buffalo hides.

THE ANASAZI

The Anasazi ("Ancient Ones") first appeared in Arizona about 2,500 years ago. The first stage in Anasazi development was the Basket Maker Period, named because of their mastery of weaving food containers, sandals, and other goods from straw, vines, rushes and yucca.

The early Anasazi lived in northeast Arizona in rock shelters and were hunter-gatherers.

By 300 A.D. the first pit houses appeared. These were semi-permanent, rounded and domed structures which were erected over shallow depressions, built with horizontal mud-chinked logs. By 400 A.D. the Anasazi had established a site near the north rim of the Grand Canyon.

Subsistence was a mixture of hunting, gathering and horticulture in the floodplains of upland and lowland drainages. Sometime during this period corn became an important crop along with beans, squash and sunflowers. Long distance trade routes were developed, evidenced by the abundance of marine shells from the Pacific Ocean and the Gulf of California. It is likely these early routes crossed the Colorado River near present-day Needles and/or Bullhead City,

Pit House — A semi-subterranean structure several feet in the ground, with a log frame and a roof of sapplings, reeds and mud. From Waldman, 1985.

heading for the Mojave River to the west, or following the Lower Colorado River southward to the Gulf of California. Parts of these trails can still be seen today.

One of the few sites in northwest Arizona is the Willow Beach Site, dated at A.D. 500, located about 15 miles south of Hoover Dam on the Arizona side of the Colorado River.

Between A.D. 500 and A.D. 1200 the growing Western Anasazi populations expanded outward to attain their greatest geographic distribution, which included the Grand Canyon area. Many of the Grand Canyon's Indian trails were developed during this period. Over 2,000 archaeological sites, including rock granaries, have been found.

44

PHASE	TIME	TRADITION		STAGE
		VIRGIN	KAYENTA	
	A.D. 1600			
	1500			
Aggregation	1400			
	1300		Abandoned	
Reorganization			Pueblo III	Long distance trade. More farming.
	1200	Abandoned	Transition	Less trade. Expansion into Grand Canyon area.
Differentiation	1100	Mesa House	Pueblo II	Small animal hunting. Farming oriented. Increased and growing population.
	1000			
Expansion	900	Lost City	Pueblo I	
	A.D. 800			Continued large game hunting. Bow replaced atlatl. Heavy reliance on hunting.
	700		B.M. III	
	600			Pit houses and storage facilities. Low and localized population-farming.
Initiation	500	First Recognized		Trade outside area. Pottery well developed.
	400			Gathering, hunting and horticulture subsistence.
	300		Basket Maker II	Pithouses.
	200			
	100			
	A.D. 1			Rockshelters. First pottery.

Stages and phases for the Western Anasazi.   Modified from Cordell, 1989.

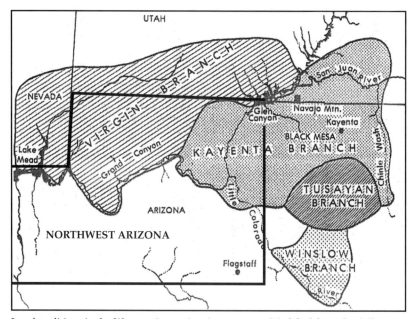

Local traditions in the Western Anasazi region.   Modified from Cordell, 1989.

By 550 A.D. the Anasazi population remained sparse and localized. Farming became more important, although there was a continued reliance on hunting large animals. In the next 300 years the bow replaced the atlatl as the primary projectile delivery implement. Inhabitants maintained wide contacts within the area and with groups in other parts of the Colorado Plateau.

Marine shells continued to be imported, indicating long-distance trade relationships. Stone slab-sided pithouses and associated storage facilities began to be integrated into larger villages. Pottery was a well-developed art.

Lost City, just north of the Colorado River in southern Nevada, with numerous contiguous surface units, was established about A.D. 700 and occupied for more than 300 years.

The population increased, leveled off and began to decline around A.D. 1100. Farming remained the basis of subsistence with an emphasis on small animal exploitation. Outside trading abruptly fell off.

Then sometime between A.D. 1150 and 1300 Anasazi occupation in northwest Arizona came to an abrupt end. The reasons are unknown, but are likely related to a prolonged drought between 1276 and 1299.

During this time in northeast Arizona and northwest New Mexico, the Anasazi became cliff dwellers. They moved from the valleys and plateaus to the safety of cliff houses for defensive purposes, protecting their families from their neighbors when food was scarce. It is thought that the Anasazi migrated southward and joined other closely related groups.

FIRST CONTACT

The first contact with Spanish explorers was in the early 1600s with continued contact throughout the 1700s. As the Spanish colonies moved northward out of Mexico into what is now New Mexico, their dominant mode of land transportation was the horse. Although Spanish officials outlawed the trafficking of horses to the Indians, they could not prevent the gradual dispersion of the animals to the tribes with whom they came in contact. It is not known exactly when the modern tribes of northwest Arizona first obtained horses.

The Spanish also introduced fruit trees, including peach, apricot and fig, into New Mexico and California, which spread into northwest Arizona via the trade routes.

THE NAVAJO

About A.D. 1050 the Navajo migrated from what is now western Canada, establishing a homeland between the Rio Grande, the San Juan and the Colorado rivers.

Sometime between A.D. 1700 and 1800 some of the Navajo migrated westward from the four-corners area to just west of the Little Colorado River (approximate western boundary of the present Navajo Reservation). They lived in log and mud hogans which can be seen today along U.S. 89 north of Flagstaff.

Hogan — A Navajo dwelling traditionally facing east, conical, hexagonal, or octagoal in shape, with a log and stick frame covered with mud, sod, adobe and sometimes stone. From Waldman, 1985.

The Navajo obtained sheep through raids against the Spanish in New Mexico in the 1700s. They slowly built up their herds, becoming master sheepherders. In the 1860s and early 1870s some of the Navajo conducted raids on Mormon settlers and communities in northwest Arizona. There was peace after the Navajo War of 1873. Today the major Navajo population center in northwest Arizona is Tuba City.

THE SOUTHERN PAIUTE

Sometime after A.D. 1300, after the Anasazi vacated the northern part of northwest Arizona, families of the Southern Paiute migrated into southwest Utah and northwest Arizona. As with the other tribes to the south, the Paiute were essentially hunter-gatherers and farmed the floodplains. One of their sites was located at what is now Pipe Spring.

Mormon settlers reached southern Utah and northwestern Arizona in the early 1860s making contact with the Paiute. During the 1860s the Paiute made raids on pioneer settlements. To protect the local settlers, forts were built at Kanab in 1864, abandoned and rebuilt again in 1868, and at Pipe Springs in 1872.

A forced peace followed and the Paiute began to work for the settlers and ranchers.

Up to the 1900s many Paiute continued to hunt, gather and farm when they could. To support their families they also sold crafts, worked as ranch and farm hands, and as domestics.

There are ten modern Southern Paiute groups of which the Kaibab is one. The Kaibab Paiute have a reservation on the "Arizona Strip" north of the Grand Canyon (see map, page 50).

THE 1800s

During the 1820s Anglo-American trappers and fur traders, including Jedediah Smith in 1826, began to travel through northwest Arizona making contact with the Pai tribes.

All the Pai tribes had established their own territories over hundreds of years. They had a hunter-gatherer, farming lifestyle where they farmed in the river and stream floodplain channels in the spring and summer and wintered on the plateaus and foothills in the fall and winter.

This lifestyle continued until the 1850s and 1860s when the white prospectors, missionaries, ranchers and settlers entered northwest Arizona and began pushing the Pai out of their way.

In the early 1850s U.S. Army explorations through northwest Arizona were seeking wagon and railroad routes to the west coast. They also experienced scattered conflicts with the Indians.

Lt. Beale's road-building expedition in 1857-1858 opened the country to Anglo settlers. In spite of uncivil behavior attributed to the tribes, the Indians were generally of a peaceful and generous demeanor. Confrontation was the result of provocation, attacks, or unjustified imposition upon the tribes. This led to the establishment of an Army garrison at Fort Mohave in 1859 and there was a short period of enforced peace. By the mid-1860s large numbers of prospectors rushed into northwest Arizona and many of the Hualapai and Havasupai worked in the white-owned mines, becoming a reliable inexpensive labor force. New frictions developed resulting in the establishment of the Colorado River Reservation south of today's Parker Dam in 1865, and in 1870 the Fort Mohave Reservation which covered a small part of California, Arizona and Nevada where they come together along the Colorado River.

The Havasupai were separated from the other Pai bands by the Hualapai War (1866-1869). They were recognized as a separate people in 1880 and received a 518-acre reservation at the bottom of Havasu Canyon.

The Hualapai were not always happy with the explorative practices of the Anglo newcomers and fought a number of wars with them. In 1874 the Office of Indian Affairs decided that the army would forcefully move the Hualapai to the Colorado River Indian Reservation where they stayed for two years until they agreed to make peace with the settlers. Many went back to work in the mines. In 1883 their present reservation was established consisting of some 997,045 acres of deep canyons and high plateaus.

By 1874 the conflicts were forcibly ended, after which the ranchers and miners colonized the habitable areas, took over the springs and started to herd cattle over large tracts of grassland. Many of the Indians of northwest Arizona took jobs in the mines, on ranches and with the railroad.

During this time period between 1850 and 1883 the Hualapai suffered epidemic diseases, including smallpox, whooping cough, gonorrhea and syphilis.

The Yavapai and Western Apache shared a common border in central Arizona and there was some intermarriage and sharing of cultural traditions. When settlers and miners entered central Arizona after the Civil War there were raids and encounters with the U.S. Cavalry, which ended with a massacre of Yavapai in December 1872. In 1875 the Yavapai were removed to the San Carlos Apache Reservation and later, about 1900, were allowed to return to central Arizona.

THE 1900s

Havasupai

At the turn of the century many Havasupai were working for miners, ranchers and the Park Service at the Grand Canyon. They also had small gardens and cattle herds on private land and land leased from the U.S. Government. By 1939, the tribe had essentially become wage earners, dependent on the Bureau of Indian Affairs (BIA) and the Park Service for jobs.

In the mid-1950s the Havasupai were restricted to a 500-acre reservation in Cataract Canyon within the Grand Canyon. A majority of the tribe lived and worked off the reservation.

In a long battle to regain some of their ancestral land, the Havasupai prevailed and in 1974, by Congressional enactment, a 160,000-acre reservation was established and 95,000 acres of Grand Canyon National Park was allocated for their permanent use. The Havasupai Reservation base is at Supai along Cataract Creek and Havasu canyons near the south rim of the western part of the Grand Canyon.

In the 1990s the Havasupai community had a population of 565 people.

Hualapai (Walapai)

There were few Hualapai on their reservation at the turn of the century. Most worked in towns, on the railroad, or with miners or ranchers. During the 1930s many Hualapai returned to the reservation and found employment with the Civilian Conservation Corps (CCC) building roads and making other improvements. When the CCC program was terminated, most chose to remain on the reservation to tend small herds of cattle. The Indian Reorganization Act of 1934 restored the powers and sanctions of tribal governments. By 1960 some 350 of the 700 Hualapai lived permanently in the reservation town of Peach Springs. During the 1960s there was chronic unemployment but this changed for the better in the 1970s with an award of nearly $3 million by the

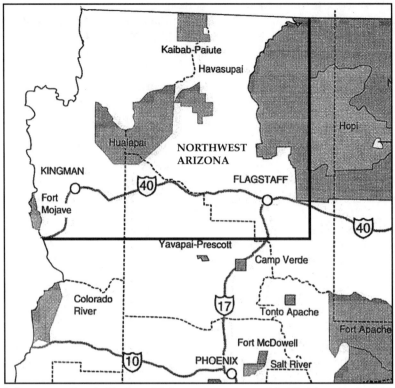

Indian reservations in Northwest Arizona.   Modified from Davis, 1996.

Indian Claims Commission for lands appropriated by the United States in 1883. The money helped develop a trading post and grocery store and established a high-quality cattle herd on the reservation

Yavapai

Two reservations were established for the Yavapai in central Arizona at former military posts in 1903 and 1910, and a third in 1935. In 1990 the Yavapai numbered about 820.

Mojave

On February 2, 1911 the U.S. Government gave the Fort Mohave tribe the status of a sovereign nation. Agriculture had been a major tribal business with cotton as a main crop. There are more than 184 homes on the reservation and a new casino. Tribal enrollment is about 1,000 citizens, many who work in Laughlin, Bullhead City, Needles and the surrounding area.

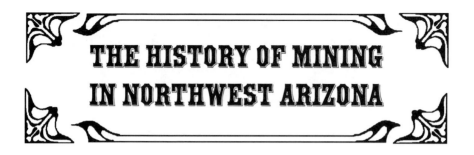

THE HISTORY OF MINING IN NORTHWEST ARIZONA

The earliest miners in northwest Arizona may have been the Aztecs whose primitive mining tools which were used to obtain turquoise from one of the largest deposits in the world have been found in the Mineral Park area. Later came Native Americans who also mined turquoise as well as salt and red clay for trading, healing, spiritual and domestic uses.

Mining in the 1800s played an important part in the economic development of northwest Arizona, and its early history tells of the harsh conditions that had to be overcome to be successful. The scorching desert heat, hostile Indians and isolation quickly separated the "men from the boys." The prospectors were men with dreams, but for only a very few would their visions of striking it rich come true. The others would help build the West in different ways.

This chapter first defines a mining district and then describes an overview of prospecting and mining, followed by a historical chronology of the early years — 1862 to 1883 and then 1883 to the present. Mining districts are then listed and described, first those districts found in the Basin and Range (western) part of the area and lastly those districts found on the Colorado Plateau.

MINING DISTRICT: A DEFINITION

A mining district designates a specific area of base and/or precious metal mineralization. It can vary in size from a few square miles to several hundred square miles and can include from one to several mining camps.

Prior to 1872 prospectors who discovered precious and base metals in a new area would get together and establish a mining district. They would form their own self-government, having agreed upon specific rules for the district that included in part how mining claims would be located and recorded, the claim size, how the claim corners would be marked in the field, and how disputes would be resolved.

The federal Mining Law of 1872 standardized many of the questions the early prospectors had to decide upon, district by district.

THE NATURE OF PROSPECTING AND MINING
IN THE SOUTHWESTERN UNITED STATES:
AN OVERVIEW

Finding the "bonanza" was every prospector's dream; however, it wasn't that easy. The prospector knew that gold and silver found in quartz veins and quartz could easily be seen on the ground. He also knew that gossans, the brown and yellow iron oxide remains of pyrite (FeS_2), were an indicator of precious and base metals. The same was true of green and blue copper staining.

The day-to-day life of prospectors and miners was hard; the men were isolated and lived under harsh conditions. Some worked at other jobs and prospected on their free time; others prospected full time. Those who didn't have money for supplies were very often grubstaked by a local business owner in return for a good share of whatever was found.

Accompanied by their burros they spread out and within a few decades prospectors had walked nearly every square foot of northwest Arizona. They searched and panned every streambed and tributary channel where there was water, looking for gold, quartz and gossan material. When mineral was found the prospector followed the channel upstream to the outcrops. They would then locate their claim, sample the outcrop and even dig pits, adits or shafts to see what might open up.

Digging on an outcrop was hard work. The hole would either be a vertical pit or shaft, a horizontal tunnel (adit) or an incline. With only simple tools the prospector would hand drill holes for dynamite, blast and then shovel the rock into a bucket or wheelbarrow. When the hole got deeper the bucket would have to be hoisted to the surface using a windlass. In a tunnel, a wheelbarrow could be used and then tracks could be laid for a wooden ore car. If timber was needed to shore up the workings, trees would have to be found, sometimes miles away, felled and cut into the required size. The pits, adits and shafts we see today are evidence of a prospector's past presence.

The prospector would take the best looking rock to be assayed. Only a very few would hit pay dirt. Some would try to work their claim into a mine and others would sell their claims to promoters who would float companies to finance development of the property. Some ventures were legitimate; others were scams to get money and run.

Before the arrival of the railroad supplies had to be brought in by way of the Gulf of California and Colorado River, or freighted over 400 miles of the Mojave Desert. Costs were very high. The miners had to pay as much as two dollars a pound for black powder. Bacon cost from 35 cents to one dollar a pound, sugar was three pounds for a dollar, and flour 15 to 50 cents a pound.

In the 1860s prospecting and mining in northwest Arizona was primitive and often dangerous. Hostile Indians were a constant threat and a number of prospectors were slain. Digging was slow and all done by hand. Quartz with

Hauling supplies to the mining camps, circa 1880s.
Courtesy of the Mohave Museum of History and Arts, Kingman, Arizona.

visible gold could be crushed in arrastras, later in stamp mills, and then sluiced and panned to recover the gold.

Lead-silver ore had to be hand picked, so only the highest grade rock could be shipped for treatment. The high-grade ore was hauled by burro or wagon to the Colorado River, transferred to river steamer down to the Gulf of California, then to an ocean steamer to San Francisco, and finally by ship around the Horn to Swansea, Wales. The expense of freight and treatment was hundreds of dollars a ton. In the 1860s gold was worth $20.67 per ounce and silver about $1.35 per ounce.

In the 1870s there were very few shafts more than 150 feet deep. Ores rich in gold and silver in the oxidized zone above the water table yielded a large production well into the millions of dollars. At that time gold was still valued at $20.67 per ounce, but silver had dropped in value to $1.15 per ounce.

Stamp mills crushed the ore. Gold was recovered by amalgamation (use of mercury) and the lead-silver ore sent to smelters in San Francisco or New Mexico for treatment. All equipment and supplies had to be hauled in by wagon. Once the mines reached the water table at depths of 100 to 300 feet there were no pumps available to go deeper.

By the 1880s most of the high-grade gold-silver-lead ore was mined out above the water table. Lower grade ores were uneconomic and all but the richest mines shut down waiting for better times.

THE EARLY YEARS: 1862-1883
(before completion of the Atlantic & Pacific Railroad)

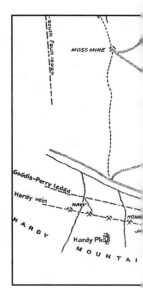

James Marshall's discovery of gold at Sutter's Mill in California in 1848 triggered the Gold Rush of 1849 to the Mother Lode country and brought thousands of fortune hunters west. In 1853 Mormon settlers in southwestern Utah discovered copper in an isolated area of what would become the Bentley Mining District. Silver and gold were discovered on the Comstock Lode in 1859 and the ensuing rush resulted in Virginia City, Nevada.

At the start of the Civil War in 1862, General J.H. Carleton and two companies of the Fourth California Volunteers were dispatched to re-garrison Fort Mohave along the Colorado River to prevent the area from falling into the hands of Confederate irregulars. Many of his men were experienced miners and prospected the hills of the Black Mountains when they had time off from garrison duty.

Prospectors followed a dry creek bed (later named Silver Creek) eastward from Fort Mohave into the Black Mountains. They established a camp along Silver Creek at a point about four miles north of present-day Oatman where surface water could usually be obtained from year-round springs. Walls of a dozen or more stone cabins built by Carleton's men were still standing in 1909. Called Silver City, the camp was in the area of Silver Creek Spring.

In 1863 John Moss, a prospector who roamed the west, found free (visible) gold in a vein (Moss Vein) outcrop about a mile north of Silver Creek. Moss is reported to have mined rich ore containing about 12,000 ounces of gold valued at $240,000 from a pocket close to the surface, which was treated in an arrastra on Silver Creek near the camp.

News of the Moss discovery brought a rush of prospectors from California and Nevada who combed the area that would soon become the San Francisco Mining District. Soon other outcropping gold-bearing quartz veins were found, including the Hardy Vein on the east side of Hardy Mountain, the Gold Dust (Victor-Virgin) about a mile southwest of present-day Oatman, and the Leland on Leland Hill (see Early Map of the Oatman Area, page 55).

The prospectors explored eastward in 1863 and silver was discovered in the Cerbat Mountains in what would become the Wallapai Mining District, north of present-day Kingman. The camp of Chloride City was established, and prospectors rushed to the area and made new discoveries seven miles to the south at Stockton Hill.

Because of hostilities with the Hualapai and Mojave Indians, early prospecting was a perilous adventure and was confined to the Silver Creek area of the Black Mountains and the Cerbat Range. Hostilities peaked in 1866 when the miners were forced to abandon the Silver Creek area for several years. Prospectors and miners at Chloride in the Cerbat Mountains were driven out

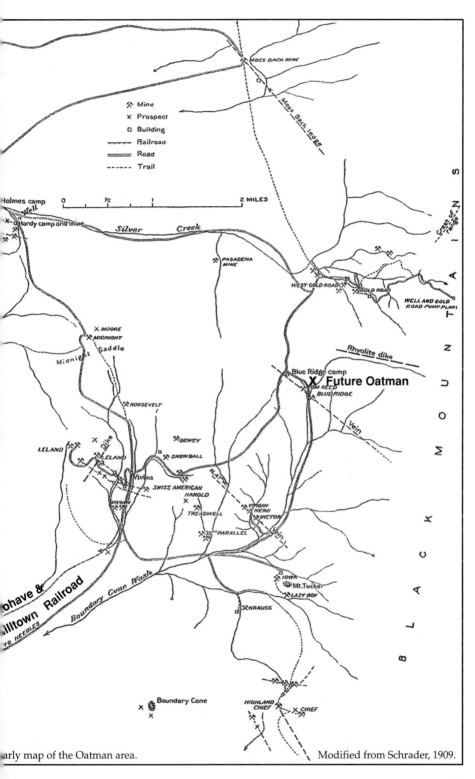

Early map of the Oatman area.   Modified from Schrader, 1909.

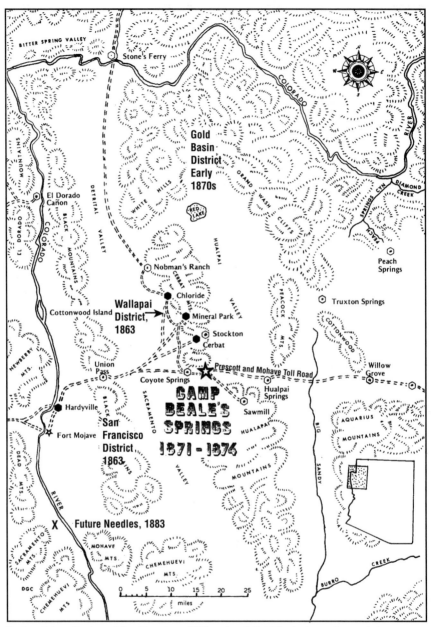

Early mining districts of San Francisco, Wallapai and Gold Basin in northwest Arizona. Kingman was established in 1883 near Camp Beale's Springs. Modified from Casebier, 1980.

several times with seven reported killed. The strongest stayed and fought the Indians, prospected and mined (see map above).

In the late 1860s an increased presence of U.S. Army personnel made life less perilous and new prospectors headed to the Black Mountains and Silver Creek.

Gold, silver and lead in fissures were discovered 10 miles south of Chloride and the camp of Cerbat was established. Gold, silver, lead and copper were discovered at Mineral Park about halfway between Chloride and Cerbat in 1870, and mining quickly began.

The Moss, Hardy, Gold Dust, Leland and Mitchell veins continued to be worked into the early 1870s, their ores treated at a small ten-stamp mill on the Colorado River near the mouth of Silver Creek about seven miles to the west. "Mineral Resources," an early 1870s government report covering the Silver Creek area, stated: "The Leland Lode runs east and west, and is about five feet in width. The gold is fine and evenly defused through the rock. A tunnel 150 feet in length follows the lode at a depth of 50 feet. Mitchell Lode also runs east and west, and the vein is from three to five feet in width. In 1871 considerable work had been done on the Leland and Mitchell lodes."

Prospectors spread out into very isolated country finding gold, silver and copper in the Gold Basin District (early 1870s), Havasu Canyon District (1873), Hackberry District (1874), Mount Trumbull District (1875), Music Mountain District and Pine Springs District (both 1880).

In the San Francisco Mining District, considerable development work was done on the Moss Vein, but no new ore was found. Other conspicuous veins, such as the Tom Reed and Gold Road, must have attracted the attention of early prospectors, although no ore was found in them until many years later. There is little record of activity in the Oatman area for nearly 30 years until 1901.

FROM 1883 TO THE PRESENT

Better times for a few of the mines came with the arrival of the Atlantic & Pacific Railroad in 1883. Supplies and equipment were easier to obtain and ore was cheaper to ship to smelters in San Francisco and New Mexico.

Individual small-scale mine operators did not have enough ore to economically ship to smelters. The Kingman Sampling Works, which measured the quantity and tested each small lot of ore to determine its value, was established in Kingman in 1883. After treatment the value of the ore would be determined using the national metal market price list for each metal less a refining fee, and the mine operator would be paid.

The price of gold held at $20.67 per ounce but the price of silver dipped below $1.00 per ounce in 1886.

In 1892 a cyanide mill and smelter furnace were built in Needles to treat gold and base metal ores from the Wallapai and other districts. In 1894 the price of silver dropped to $.64 per ounce and the mill shut down.

In 1894 the Arizona Sampling Works opened in Kingman. The two sampling businesses competed until January 2, 1901 when fire from a railroad coach on the sampler spur consumed the Kingman Sampling Works.

Kingman quickly replaced Chloride as the center of supply activity, and only the richest mines survived.

On the Colorado Plateau in the Grand Canyon area, prospectors discovered and worked small deposits of asbestos, copper and silver.

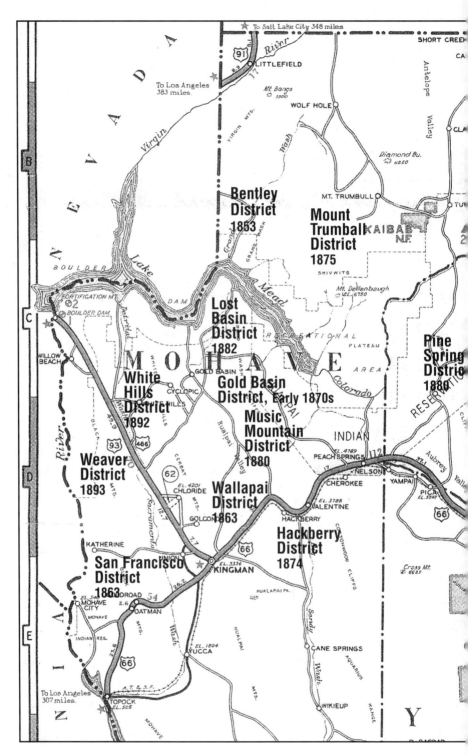

Mining districts of Northwest Arizona.

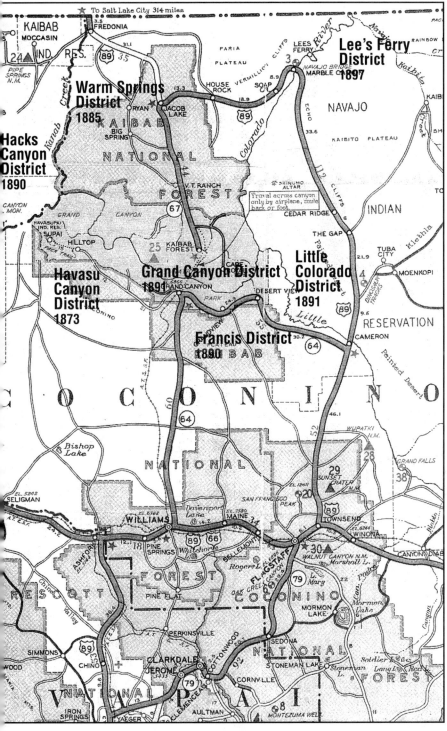

Base map is a 1939 Arizona Road Map.

After completion of the Arizona and Utah Railroad from Kingman to Chloride in 1899 mining took on new life in many of the nearby districts. Speculators invaded the area and smelters were built at Needles (1904) and Prescott (1906). Improvement in milling methods led to exploitation of complex lead-zinc ores in the Wallapai District, which reached a peak from 1915 through 1917 because of higher metal prices during World War I.

In the San Francisco District gold was discovered at Gold Road in 1900 that resulted in a rush to the area. Between 1901 and 1912 a number of rich gold veins were discovered in the Oatman area, including the Tom Reed.

When the price of gold jumped from $20.67 to $35 per ounce in 1933 gold production increased and reached a peak in 1937-38, until 1942 when Executive Order L-208 halted all nonessential mining during World War II.

In the west there was little mining after the War until the early 1960s when, in the Wallapai District, the Duval Corporation explored, developed and began large-scale mining of the Ithaca Peak porphyry molybdenum-copper property north of Kingman which operated until 1982.

On the Colorado Plateau uranium was discovered after World War II with two booms, the first in the 1950s and the second in the late 1970s. Uranium mines operated into the mid-1990s.

In the last 20 years major mining companies have looked at the mining districts in northwest Arizona without making any new economic discoveries.

MINING DISTRICTS IN THE BASIN AND RANGE PROVINCE: WESTERN PART OF NORTHWEST ARIZONA

The earliest prospecting and mining was in the mountains of the Basin and Range Province. With the exception of the Ithaca Peak porphyry molybdenum-copper deposit north of Kingman all the other mineralization occurred in veins of gold, silver, copper,

San Francisco (Oatman) and Wallapai mining districts.

60

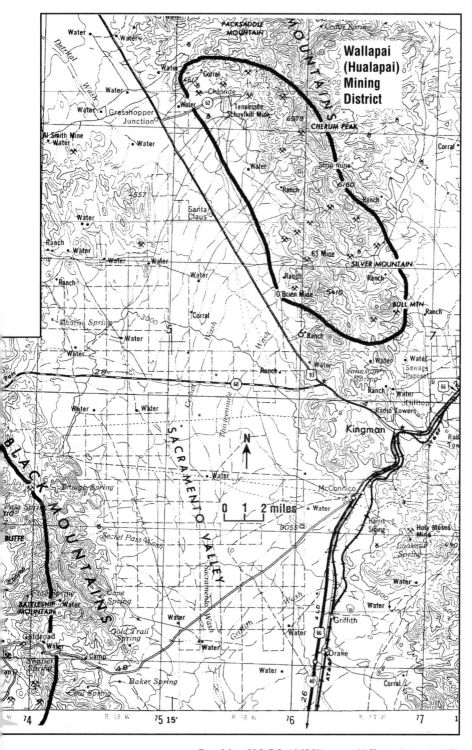

Base Map: U.S.G.S. AMS Kingman 2° Sheet, Arizona, 1972.

lead and/or zinc in pre-Cambrian granites, gneisses and schists, or in overlying Tertiary volcanics or small associated placer deposits. The two largest mining districts are the Wallapai and the San Francisco (Oatman) districts. Smaller districts include Weaver, White Hills, Gold Basin, Lost Basin, Music Mountain, and Hackberry.

Wallapai (Hualapai) Mining District

The Wallapai Mining District is in a 10-mile long, three to five mile wide northwest trending mineralized belt in the Cerbat Mountains, six to 16 miles northwest of Kingman. The district includes the camps of Chloride, Cerbat, Mineral Park, Ithaca Peak and Stockton Hill.

The Cerbat (Indian word for bighorn mountain sheep) Mountains consist largely of pre-Cambrian schist, gneiss and granite, intruded by granite-porphyry and lamprophyric dikes, and overlain in places by Tertiary volcanic rocks. Mineralization, Tertiary in age and related to the periods of widespread volcanism, mainly consists of lead and zinc with lesser gold and silver in quartz and calcite veins and a porphyry molybdenum-copper deposit. There is a classic mineral zoning in the district outward from the molybdenum-copper core at Ithaca Peak to zinc, lead and silver. Near surface ores are oxidized with a well developed secondary zone of enrichment underlain by lower grade sulphides.

Base-metal ores (copper, lead, zinc) below the water table were sparsely mined until after completion of the Arizona and Utah Railroad from Kingman to Chloride in 1899. With good water pumping equipment, lead-silver ores were mined below the water table and with subsequent improvements in milling methods, complex lead-zinc ores were mined.

Zinc-lead mining reached its peak from 1915 through 1917 owing to higher metal prices during World War I, and declined abruptly thereafter.

After World War I, exploitation was confined to high-grade gold veins. When the price of gold increased from $20.67 to $35.00 per ounce in 1933 production increased, reached its peak in 1937-1938, and was halted during World War II.

From 1950 through 1956, production of gold was less than 100 ounces per year.

Before 1904 the value of combined metals produced is estimated at $5 million. From 1904 through 1956 the mines of the district produced about 125,000 ounces of gold. The estimated value of combined metals produced in the district up to 1960 is about $10 million. The value of metals produced from Duval's Ithaca Peak mine after 1962 is in excess of $1 billion.

Chloride: The camp of Chloride City, founded in 1863, was named after the character of its rich silver ore. Mines located on prominent vein outcroppings included the Silver Hill, Golden Fleece, Tintic and Independence.

The Chloride Post Office was established in 1873 and is the oldest continuously operating post office in Arizona.

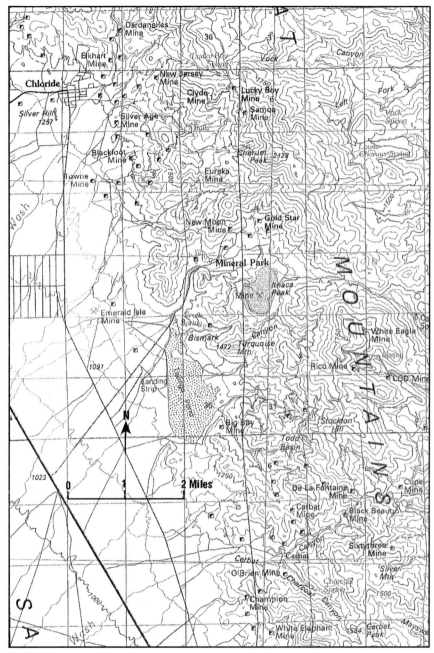

Wallapai Mining District   Map: USGS 1:100,000 Davis Dam Sheet, 1982.

Some of the ores were so rich that they were shipped to San Francisco and then to Swansea, Wales for treatment and netted their owners large rewards.

After a peace treaty with the Hualapai in 1870 there was a rush of about 1,000 prospectors to the district with 2,700 claims filed by 1872.

In Chloride there was a brewery, general store, a blacksmith and several saloons. Small silver bars molded by a local assayer worth $5, $10 or $50 were used as money.

Ore was crushed at a five-stamp mill at Mineral Park or Pioneer and a crude furnace was in operation. With the completion of the Atlantic & Pacific Railroad through Kingman in 1883 and the opening of the sampling works in Kingman, it became profitable for small operators to ship low-grade ores.

In the early 1890s Chloride boomed, and there were investors from the east and west coast and Europe.

During its heyday at the turn of the century more than 75 mines operated. Chloride was serviced by both stagecoach and the Arizona and Utah Railroad, a branch line of the Santa Fe Railroad built in 1899.

Chloride, with a population of more than 1,500 people, served as the distribution center for nearby camps, mines and other districts.

Mine production slowed by 1904 and continued to diminish by 1910.

Hundreds of people lived in Chloride during World Wars I and II.

The Tennessee Mine was one of the large producers of silver-lead ore. By 1948 production was about $7.5 million.

Today Chloride is a tourist and artist community with a population of about 300.

Stockton Hill: This was one of the oldest camps, dating to the early 1860s. High grade silver-lead, with some gold, and copper-bearing quartz veins in pre-Cambrian gneiss and schist were mined to the water level at about 100 feet from about 12 mines and shipped to Swansea, Wales. In the 1880s ore was mined below the water table, milled at Mineral Park and Cerbat and then shipped to smelters in San Francisco and New Mexico. In the early 1900s concentrates were hauled by wagon to Kingman and then by rail to the smelter at Needles. Stockton Hill is reported to have produced many millions of dollars worth of ore mainly from the Banner, Cupel and De La Fontaine mines.

Cerbat: The mining camp of Cerbat was established after gold, silver and lead in fissure veins were discovered nearby in the late 1860s. In 1871 a small smelter operated for a short time but failed for want of proper fluxes. Cerbat grew to a population of about 100 by 1873 and became the Mohave County seat in 1873. The population, including two lawyers, two doctors, a free public school, a smelting furnace, a city hall and a court house, grew to about 700 in the mid-1870s.

A better designed smelter was built to treat local ores. Operations declined in the 1880s.

The leading mines were the Golden Gem, Idaho, Flores, Esmeralda, Cerbat and Oro Plata. By 1900 these mines produced more than $2 million in high-grade ore. In 1907 a 40-ton mill was built near the Golden Gem Mine and the mine dumps were treated. Operations continued sporadically into the 1950s.

Mineral Park: The Mineral Park Camp is located between the Chloride and Cerbat camps. The first location was made in 1870 (the Mayflower), and the townsite was laid out in 1871. It was named Mineral Park because it was situated in a beautiful park-like basin between mountains rich in minerals.

In late 1871 a five-stamp mill was built and began crushing local ores. During the early years considerable gold-silver ore was mined and shipped to the Selby Smelter in San Francisco at a freight cost of $125 a ton. Some ore was freighted eastward by oxen over the old government trail to New Mexico. The five-stamp Keystone Mill was built in 1876 at a reported cost of $125,000 and, for a few years, treated most of the ore from the Cerbat Mountain region, including the Chloride, Cerbat and Stockton Hill camps.

The town of Mineral Park was quick to grow with seven saloons, a school, mercantile stores and private dwellings. For a brief time in 1877 it was the Mohave County seat.

The *Mohave County Miner* began weekly publication in November 1882. With the arrival of the A&P Railroad at Kingman in 1883 came increased mining activity and by 1884 Mineral Hill Camp had a population of more than 700, including many Indians and a sizeable Chinese section. There were two stores, a hotel, restaurant, five saloons, a blacksmith shop, school house, court house, a hospital, a red-light district and a Chinatown with opium dens and several laundries.

Mining declined in 1885 and so did the camp.

Boarding tent and workmen of the Aztec Mine at "Turquoise Mountain," near Mineral Park, 1901.
Courtesy of the Mohave Museum of History and Arts, Kingman, Arizona.

The ore deposits contained gold, silver, lead and copper in fissure veins. Important mines were the Lone Star, Keystone, Fairfield, Paymaster, Ithaca and Nighthawk. Many of the early mines could not work below the shallow water table. Mining operations declined in 1882 as the price of silver dropped, with the exception of several mines that had mills and these continued to operate for a few more years.

With the coming of the A&P Railroad in 1883 Kingman Siding became the center of activity in the area. Kingman rivaled Mineral Park in size by the late 1880s.

After the turn of the century a number of small mining operations began when new discoveries of lead, zinc and copper were made, and a few of them continued to operate as late as World War II. With depletion of the higher grade ore, lower metal prices and rising costs, all mining in the Mineral Hill area ceased except for a few turquoise operations.

In 1910 Utah Copper Company geologists noted "porphyry copper" type mineral outcrops, and drilled a churn drill hole on the southeast side of Ithaca Peak.

After World War II large low-grade copper deposits became an object of exploration by major mining companies in the southwest. In 1948 Kennecott Copper Company drilled six holes in and around Ithaca Peak.

In October 1958 Dr. Harrison A. Schmitt (father of astronaut Harrison Schmitt), a consulting geologist for Duval Corporation (a subsidiary of United Gas, Inc. and later Pennzoil Company) and a giant in the field of porphyry copper deposits, visited the area and Duval began acquiring claims in the Mineral Park area. Between 1959 and 1962 they drilled 89 churn drill holes and 34 diamond drill holes, many of which needed a helicopter to place the drill rigs. A crescent-shaped porphyry molybdenum-copper ore body was outlined beneath Ithaca Peak.

In mid-1962 an underground sampling program was begun to check the drill hole sampling consisting of 1,275 feet of adit, 1,216 feet of crosscut and raises totaling 970 feet, with excellent results. This showed a reserve of some 60 million tons of molybdenum-copper ore averaging 0.76% of copper. Sufficient water was available and the final feasibility study was completed in August. The decision to develop the Ithaca Peak property was made in October 1962, and Duval purchased the old townsite of Mineral Park.

Millions of tons of overburden were removed to expose the ore body which would be treated in a $21 million 12,000-ton per day copper concentrator and auxiliary facilities at Mineral Park. Water required for processing was 3.5 million gallons per day, pumped 14 miles from the Sacramento Valley to the west.

Production was increased to 20,000 tons per day. Lower grade copper oxide ore was placed on dumps and leached. The copper recovered paid for the costs of mining and processing and the molybdenum recovered was the profit of the operation. The mine operated until December 14, 1981.

The Mineral Park property produced 323,000 tons of copper, 250,000 tons of molybdenum and five million ounces of silver with a total value of more than $1 billion.

Cyprus Bagdad Copper Company purchased the property in 1985 and operated a solvent extraction and electrowinning process which produced pure copper cathode until 1997, when Cyprus sold the property to Equatorial Mining Company of Australia.

In the late 1960s several nearby properties were mined for copper oxides including the Emerald Isle Mine, operated by El Paso Natural Gas Company, where copper was leached from chrysocolla.

Today Equitorial has a small operation at Ithaca Peak, producing copper from the leach dumps using a solvent extraction and electrowinnings process. Turquoise is also mined on a small scale.

San Francisco (Oatman) District

The San Francisco District is in the Black Mountains east of the Colorado River and extends from a few miles south of Boundary Cone northward about 25 miles to a point north of Union Pass and Pyramid. The district ranges from six to 10 miles wide, and included the camps of Gold Road, Oatman, Silver Creek, Union Pass, Pyramid and Katherine.

Later, important gold discoveries in the area of present-day Oatman led to designation of that general area as the Oatman Mining District.

Oatman: The Moss Vein was discovered near Silver Creek in 1863 and created the first rush to the area (see The Early Years, page 54).

Sometime prior to 1900 two prospectors discovered and staked a claim which became known as the Snowball Mine (see map, page 55). A small camp called Vivian grew along with a miner's union known as Snowball. It is reported that at the turn of the century (1900) there was a Chinese restaurant, two saloons and a number of families living in the camp.

In 1900 rich ore was found in the Gold Road vein and created a small gold rush to the area. Between 1901 and 1912 a number of discoveries were made along earlier worked veins and new veins, including the Tom Reed.

Some of the gold ore from the earlier mines in the western part of the district was shipped and treated at the Leonora Mill at Hardyville along the Colorado River in 1901-02 (near present-day Bullhead City).

In 1903 the Mohave Gold Mining Company, before having proven ore reserves in the Leland Mine, built 17 miles of narrow-gauge railroad from their mine to a point on the Colorado River opposite Needles. They also erected a mill at a site called Milltown along their railroad about 11 miles southwest of the mine. About 4,500 tons of gold ore valued at $40,000 was mined and shipped to the mill. Late in 1904 the mill was closed, and by 1907 many miles of the railroad were washed out by floods.

At Gold Road, along the Tom Reed Vein and at other veins in the district, ore bodies were discovered, developed and mined. Mining properties were promoted, bought and sold.

In 1904 a 10-stamp mill was built at the Tom Reed Mine.

The "Mining News" of 1912 referred to the camp of Oatman as having been established between the Tom Reed and the United Eastern mines.

Problems at Gold Road (Note the man and broken front wheel), circa 1915.
Courtesy of the Mohave Museum of History and Arts, Kingman, Arizona.

Access to the district was by wagon road over Sitgreaves Pass to the railhead at Kingman to the east, and west down Silver Creek to the steamer landing at Hardyville or down Boundary Cone Wash along the Mohave and Milltown Railroad to Milltown and the railhead at Needles.

In 1914 the wagon road from Kingman to Gold Road, Oatman and on to Needles was designated as the National Old Trails Highway.

The same year the newly formed United Eastern Mining Company discovered buried gold ore just north of Oatman along the Tom Reed Vein and a 200-ton mill was built, later enlarged to 300 tons.

The year 1915 was a boom time for the Oatman District after word got out about United Eastern Mining Company's $6 million ore body. Between 1915 and 1922 a number of buried ore bodies were discovered along or near the Tom Reed Vein. During this time scores of shafts were sunk all over the district to depths of 300 to 500 feet seeking buried ore bodies, most without success.

In 1916 south of the Tom Reed Mine along the Tom Reed Vein complex, new buried ore bodies were discovered in the area of the Big Jim, Grey Eagle, and Black Eagle claims.

In 1917 the United Eastern Company purchased the Big Jim Mine and in 1919 the Tom Reed Gold Mine Company sued United Eastern claiming that the Big Jim ore body was the faulted segment of their Grey Eagle ore body and that the apex law should apply. The litigation was decided in favor of United Eastern in 1921.

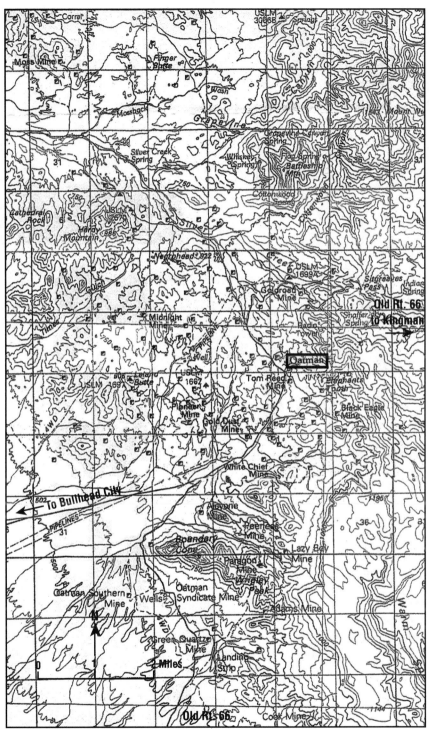

Central Oatman Mining District.
Map Base: USGS 1:100,000 Davis Dam (1982) Needles (1983) Sheets.

TABLE 1
INDIVIDUAL MINING COMPANY PRODUCTION
OATMAN MINING DISTRICT, MOHAVE COUNTY, ARIZONA

Mine	Gold $20.67/oz 1897-1933 Tons of Ore	Mined Grade oz/T	Gold $35/oz 1934-1942 Tons of Ore	Mined Grade oz/T	Total Tons of Ore	Average Mined Grade oz/T	Total Value
Tom Reed Mining Co.	981,090	0.70	205,125	0.32	1,186,215	0.64	$16,492,791
United Eastern Mining Co.	687,038	1.12	0	0	687,038	1.12	$15,905,200
Gold Road Mining Co.	737,926	0.47	775,895	0.22	1,513,823	0.32	$13,143,260
Total Production	2,406,054	0.74	981,020	024	3,287,076	0.59	$45,541,260

From Durning, 1984.

To the end of 1921 $3,952,700 in dividends were paid.

By 1924 the United Eastern ore body was mined out and the United Eastern-Big Jim mines closed. Diamond-drill exploration on the property was unsuccessful. In 1931 the Big Jim Mines Company reacquired their old property by lease.

In 1933 the price of gold increased from $20 to $35 per ounce, and the District was revived resulting in the reopening of the Tom Reed and Gold Road Mining Company properties. The Big Jim operation also went into production. An aerial tramway was built to haul their gold ore to the United Eastern 300-ton cyanide plant for treatment.

All operations in the Oatman District were shut down in 1942 as the result of Executive Order L-208.

After the war sporadic attempts were made to explore and mine properties in the Oatman District with little success.

Recent drilling exploration at the United Eastern Mine indicate ore reserves of 200,000 tons averaging 0.2 oz. Au/ton.

Union Pass (Katherine) Area: The Union Pass area is located in the north part of the San Francisco (Oatman) District from Union Pass westward to the Colorado River (near Davis Dam).

Gold and silver-bearing quartz veins were discovered in the 1860s, yielding more than $2 million (about 85% in gold). The veins occurred in pre-Cambrian granite and gneiss overlain with remnants of Tertiary volcanics. The ore deposits are similar to those found near Oatman.

The largest producer was the Katherine Mine (originally called the Catherine Mine) which was discovered in 1900. The road used to haul ore from the Sheeptrail Mine to the mill near the Colorado River passed near an isolated granite knob on the flat gravel plain. In September 1900 teamster S.C. Boggs panned some gold from the granite outcrop and mined about 2,000 tons of ore which was processed at the Sheeptrail Mill between 1900 and 1903. The mine was leased in 1903, then sold in 1907 and operated until 1909. In 1919 there was renewed development in the Katherine Mine, including a new shaft sunk to 950 feet and mining began again.

The renewed activity resulted in the establishment of the Catherine townsite about two miles east of the mine. A 75 tons-per-day cyanide mill was completed in June 1925, and later enlarged to treat 250 tons per day. Catherine townsite was a boom town.

The mill operated to 1929 and then periodically through the depression years due to the fluctuating metal prices and a fire which damaged mine structures in 1934. The mill shut down in April 1934 and mine operation stopped in 1942.

The Katherine Mine produced less than $2 million in gold and silver between 1900 and 1940. The mill, processing ore from other area mines, produced $3.6 million in gold and silver between 1900 and 1942.

Other mines in the area included the Roadside, Arabian, Tyro, Sheeptrail-Boulevard, Frisco, Pyramid, Golden Cycle and Black Dyke Group.

The Tryo Mine has estimated reserves of 300,000 tons averaging 0.1 oz. Au/ton and the Frisco Mine has 20,00 tons averaging 0.2 oz. Au/ton.

Weaver District

The Weaver Mining District is north of the San Francisco Mining District in the northern Black Mountains about 10-15 miles west and northwest of Chloride. The camps of Gold Bug, Mocking Bird, Pyramid and Pilgrim are on the eastern slope of the Black Mountains and Virginia Camp is on the western slope.

Gold in quartz veins in pre-Cambrian granites, gneisses and volcanics was discovered as early as 1893 at Gold Bug several miles north of Weaver and in 1903 near the Pilgrim Camp. Properties were developed by numerous shafts up to 300 feet deep, and early ores from the district were treated at two stamp mills, one on the Colorado River. Some 1,900 ounces of gold were produced before 1932. The greatest activity was between 1932 and 1942 with a total production to 1960 of about 63,200 ounces of gold.

Gold Basin District

Gold Basin is located in the eastern part of the White Hills about 60 miles north of Kingman. Gold was discovered in the early 1870s, but development was inhibited by the remoteness of the area and the scarcity of fuel and water. In the early 1880s a five-stamp mill at Grass Springs treated ore consisting of gold associated with lead and copper from veins cutting pre-Cambrian granite and schist. About $100,000 in gold was mined from the Eldorado Mine before 1900.

At the turn of the century a 10-stamp mill was built and treated gold ore from the surrounding mines until it was destroyed by fire in 1906. Because of the scarcity of timber and the isolated location, there was only small scale production to 1920, followed by a period of inactivity from 1920 to 1932. A few mines were reopened from 1932 to 1942 and the district has been essentially dormant since. Total production from the district was about 15,000 ounces of gold.

Lost Basin District

Lost Basin is located just south of the Colorado River (near present-day Lake Mead). Gold-quartz veins in pre-Cambrian and schistose rocks were discovered in 1882, yielding a small production which was treated at Grass Springs or in small arrastras. A post office operated between 1884 and 1891. There were small placer operations in the early 1930s.

White Hills District

The White Hills District covered a mineralized area about two miles in diameter, located about 28 miles north of Chloride and 14 miles southwest of the Gold Basin District. Hualapai Indians frequented this area for hundreds of years to obtain red iron oxide to adorn their faces. In 1892 an Indian showed a prospector at Gold Basin a piece of rich silver ore and led him to its source in the White Hills.

Silver in gold-bearing quartz veins cut pre-Cambrian gneissoid granite. A rush followed and the White Hills Camp reached its peak in 1894 with a population of about 1,200.

In 1895 the main property, the Hidden Treasure Mine, and adjacent area were sold to an English company for $1,500,000. The company's development of the property included a fully equipped 40-stamp mill, an elaborate water-supply system which included seven miles of seven-inch wooden pipe line, and a large reservoir at a cost of more than $150,000.

A small town was laid out with water, electric lights and fire hydrants on the streets. The town included stores, offices, dwellings, a church, schoolhouse, a hotel and a number of saloons.

Mines reached a depth of nearly 1,000 feet with the water table standing at 400 to 600 feet in the mines. The water from the mines was of good quality for use by the people and for milling purposes.

Coal, Joshua trees and yucca palms were used to heat boilers for the steam engines.

Stagecoach at 4th and Front streets, Kingman leaving for Chloride and White Hills, circa 1908. Courtesy of the Mohave Museum of History and Arts, Kingman, Arizona.

In 1900 the company went broke and most of the people left.

The district was revived in 1905 when a 10-stamp mill was built and mine dumps were successfully worked.

By 1909 there was a store and a post office. A stage provided passenger and mail service with Chloride and Eldorado Canyon, Nevada.

To 1909 the district produced about $3 million, mainly from silver from 15 mines.

Music Mountain District

Music Mountain is located in the foothills of the Grand Wash Cliffs. Pre-Cambrian granite, schist and gneiss intruded by dikes of diabase and granite porphyry and gold-quartz veins are overlain by Paleozoic limestones of the Colorado Plateau to the east.

Discovered about 1880, the high- grade gold ore shoots, only a few inches wide, yielded about $20,000 in bullion prior to 1904 with only a small production since. The main mine was the Ellen Jane.

In the 1930s bat guano discovered in a cave (Bat Cave Mine) was mined using a cable tram until 1960.

Since the mid-1890s limestone has been quarried from the Nelson Mine, five miles east of Peach Springs, by the Grand Canyon Lime and Cement Company.

Hackberry and Cottonwood Districts

The Hackberry District is located in the Peacock Mountains south of the town of Hackberry near U.S. Route 66. In the spring of 1874 four men, whose horses were stolen by Indians, found spring water near a hackberry tree. Nearby they found a rich silver ledge and located several mining claims.

In the summer of 1875 gold was discovered eight miles east of Hackberry in the hills of Cottonwood Creek and the Cottonwood District was established. Gold-bearing quartz veins occurred in pre-Cambrian schist.

A five-stamp mill was erected at the Hackberry Mine in 1876, only to be destroyed by fire in 1877. A new 10-stamp mill was built and by the end of the year, 222 bars of silver were produced valued at about $223,000.

A camp was established about a mile and a half northeast of the mine and grew in population to over 400. Buildings included a post office, saloons, a restaurant, a general store and school. The Hackberry Mine was opened by three inclined shafts, one 425 feet deep.

In 1882 a new gold strike in Cottonwood brought additional men to the area and several mines were developed that yielded a small production until about 1934. Total production was about $1 million.

When the A&P Railroad arrived a mile to the north, the town of New Hackberry was established around the railroad station and became a shipping center to supply the mining camps which included Gold Basin, Lost Basin and Music Mountain.

MINING DISTRICTS ON THE COLORADO PLATEAU : GRAND CANYON AREA

During the 1870s local Indians told of potential mineral riches within the Grand Canyon gorge. Before the completion of the Atlantic & Pacific Railroad in 1883, people coming to the Grand Canyon area traveled by horse, wagon or stageline. Prospectors and several mining companies came to the canyon to investigate showings of copper, lead, asbestos and other deposits. Hundreds of claims were staked under the Mining Law of 1872, but most of the deposits were found to be marginal at best.

Access into the canyon was extremely difficult. Much time and effort was spent building and improving trails which could be used with burros to haul supplies in and ore out.

Several types of mineralization were found in the Grand Canyon area. Very low-grade placer gold was discovered in Colorado River gravels upstream from Lee's Ferry

Asbestos was found in the lower gorge of the Grand Canyon in the 1880s. Asbestos veins occurred along contact metamorphic zones where black diabase dikes and sills intruded the Bass Limestone some 900 million years ago. Small quantities of asbestos were mined from the Bass, Hance and Thunder River mines.

Copper was found with lead and silver in fissure veins in the Vishnu Schist in Copper Creek near Bass Camp. Bass hauled enough ore on burros to fill two railroad cars.

Copper with minor lead and silver and later uranium was also found in breccia pipes along Cataract/Havasu Creek in the western Grand Canyon, at the Orphan Mine near Grand Canyon Village, and in the Hacks Canyon District.

Uranium was discovered on the Colorado Plateau in the mid-1940s. There were two uranium booms in the Grand Canyon area. The first was in the 1950s with discovery of uranium oxide in the Hacks Canyon Mine and at the Orphan Mine. The second was in the late 1970s when skyrocketing prices encouraged exploration for mineralized breccia pipes. There was a frenzy of exploration in the southern part of the Colorado Plateau in the late 1970s and early 1980s, and thousands of mining claims were staked.

Uranium pockets were also found associated with organic material in the Shinarump Conglomerate and overlying Petrified Forest Member near Cameron and Lee's Ferry, Arizona resulting in many small open pit mines.

Many breccia pipes had small showings of uranium and were not economic. Declining prices since 1979 and environmental battles have deterred development of some economic deposits.

The Canyon Mine, owned by Energy Fuels Nuclear, about 12 miles southeast of Grand Canyon Village near Red Butte was one of the few mines near the Grand Canyon that operated into the 1990s.

A number of discoveries of uranium-bearing breccia pipes were made in the early 1980s on the Kanab Plateau and worked into the early and mid-1990s. They included the Pigeon Mine on the Arizona Strip, the Her-

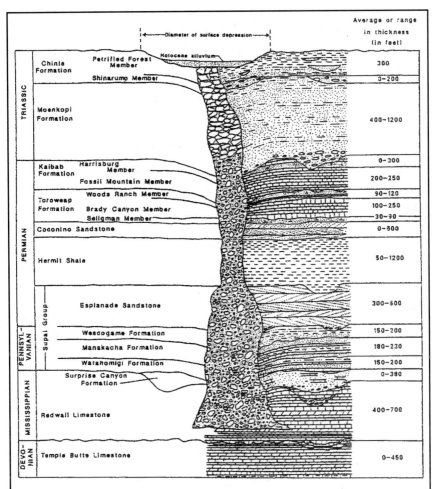

			Average or range in thickness (in feet)

General Structure of a Breccia Pipe (from Billingsley, 1997).

Breccia pipes in the Grand Canyon area are only found between the Redwall Limestone and the plateau surface. They formed beginning as a cave where water dissolved the limestone of the Redwall limestone and the roof then collapsed, finally reaching the surface, which now appears as a depression. Many breccia pipes are not mineralized. Those that are mineralized have copper, silver and uranium deposited during two time periods: the first was about 260 million years ago and the second about 200 million years ago. Mineralization from circulating waters filled in the open spaces cementing most of the brecciated rock. Only a few of the mineralized breccia pipes contained economic mineral deposits and appear to be scattered randomly.

mit Mine, the Pinewood Mine, and the Arizona 1 Pipe, about five miles southwest of the Hacks Canyon Mine.

There has been sporadic oil exploration in the Grand Canyon Area. The first wildcat well was drilled on the north rim in 1906. Since then numerous wildcat holes have been drilled north and south of the Grand Canyon with little more than slight oil stains.

The Grand Canyon area contains a number of mining districts including Bentley, Centennial, Mount Trumbull, Pine Springs, Havasu Canyon, Hacks Canyon, Warm Springs, Francis, Little Colorado River and Grand Canyon. Only the Bentley, Havasu Canyon, Hacks Canyon, Mount Trumball, Warm Springs and Grand Canyon districts will be discussed here.

Grand Canyon District

The Orphan Mine: Copper was discovered by Daniel Hogan and Henry Ward in 1891 in the Coconino Sandstone, 1,100 feet below the south rim near today's Maricopa Point. A few years later Hogan located a mining claim over his discovery called the Orphan Mine. Access to his discovery, called "Hogan's Slide," was by way of ropes, ladders and rock steps. In 1906, after a little mining, Hogan and Charles Babbitt were issued a mining patent for 20.64 acres, covering his ore deposit and about four acres along the rim.

In the 1930s Hogan built 20 cabins, the Grand Canyon Trading Post (Kachina Lodge), curio shops, and a saloon on his four acres.

In 1946 Hogan sold the Orphan Mine property for $55,000 to a Mrs. Jacobs of Prescott who wanted to use it for tourist potential.

Over the years the park developed around the patented mining claim and in 1951 prospectors found uranium ore on Hogan's claim.

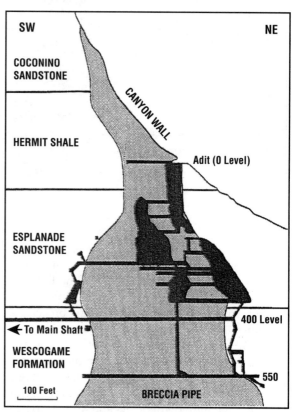

Profile of the Orphan Mine, looking northwest. Black areas are stopes; level 400 extends to the southwest and the main shaft to the surface.   Modified from Billingsley, 1997.

In 1956 an aerial tramway was installed between the mine and the rim, each bucket holding 800 pounds of ore. In 1959 a new vertical shaft was completed to a depth of 1,590 feet, which increased production from 1,000 tons per month to 9,000 tons per month. During this period rich uranium ore was mined underground and hauled to Tuba City for processing.

Between 1951 and 1969 the Orphan Mine produced more than 100,000 tons containing at least 0.1% U_3O_8, 4.3 million pounds of copper, 107,000 ounces of silver, and 3,283 pounds of vanadium oxide.

The ore deposit occurred in a breccia pipe in the Redwall Limestone and included uraninite with copper, silver, lead, zinc, cobalt, nickel, molybdenum and vanadium.

In 1962 President Kennedy signed legislation transferring ownership of the patented mining claim to the National Park Service, allowing underground mining until 1987. The Orphan Mine headframe, located between Maricopa and Powell Points, has been a landmark for many years. The Orphan Mine ranked among the five most important deposits of uranium in the United States.

Bass Camp: In 1890 William Wallace Bass discovered asbestos and later copper in a contact metamorphic zone located along the contact where a black diabase sill intruded the Bass Limestone. This was near the bottom of the Grand Canyon about 20 miles west of today's Grand Canyon Village (see Grand Canyon map, page 96-97). Over the next 20 years a small tonnage of asbestos and copper was mined. In 1926 all of the Bass properties were acquired by the Santa Fe Land Development Company

Hance Camp: John Hance prospected the eastern Grand Canyon area between 1883 and 1890, discovering asbestos veins where a black diabase sill intruded the Bass Limestone about 12 miles east of today's Grand Canyon Village (see Grand Canyon map, page 96-97). Claims were located over the asbestos veins in the early 1890s and 16 claims were patented by the Hance Asbestos Company in 1901. After a small production mining ceased in 1902, and sometime thereafter the claims were sold for taxes. They were finally owned by the Hearst family.

In the 1970s the last of the Hance claims were in negotiation for reversion to the National Park Service.

Havasu Canyon District

This district covers the Cataract Canyon and Havasu Canyon areas near Supai. Silver and lead mineralization in the Redwall Limestone below Bridal Veil (Havasu) Falls was discovered in 1873 and mining of the Moqui Quartz claim began, reaching a peak in the late 1870s. Every tributary along Cataract/Havasu Creek was explored by prospectors and many gossans and breccias were evaluated. In 1883 claims were filed in Upper Carbonate Canyon where lead-silver, gold, and copper mineralization was found in a breccia pipe.

Numerous claims were filed in the late 1800s on showings of lead, silver and copper. Over the years claims were bought and sold and new companies organized to raise capital to develop the properties. There were some promotional scams of gold and platinum discoveries.

In the 1920s vanadium was recognized in the area of the 1873 discovery.

Because of the remoteness of the area where mineralization was found, sheer cliffs accessible by narrow paths, iron scaffolding over 250 feet high and wooden ladders, only handpicked samples for testing were obtained.

Mining operations in other areas of Cataract Canyon began to decline in the early 1900s.

During World War II new attempts were made to produce lead and vanadium from the old claims. About 156 tons of handpicked high-grade oxidized lead-silver ore was packed on ponies up the 11 mile trail to the rim and then trucked to the smelter in El Paso, Texas. The high cost of mining and transportation from the remote locale doomed the operation.

The claims were purchased by the National Park Service in 1957. In 1975 the mines, canyon and adjacent plateau were returned to the Havasupai who do not permit access to any of the mines within their reservation.

Occasionally in recent years backpackers have found caches of early miner's and prospector's tools and supplies. Campers have dumped tons of trash into the tunnels and shafts.

Hacks Canyon District

The Hacks Canyon District is located on the Kanab Plateau north of the north rim of the Grand Canyon. The discovery of a very small amount of gold along Kanab Creek resulted in a five-month "Kanab Gold Rush" in early 1872. In 1890 copper mineralization was discovered in a breccia pipe in Hacks Canyon and by 1900 it had become a center of mining activity with minor production.

In 1945 uranium was discovered in mine dumps and between 1948 to 1953 copper and uranium were mined. Various mining companies leased properties and explored for uranium in the Hacks Canyon area in the late 1950s, 1960s and early 1970s when mining claims covered the canyon and surrounding areas of the Kanab Plateau.

In 1973 the Hacks Canyon mining property was acquired by Western Nuclear which discovered two new mineralized breccia pipes. The three uranium-copper breccia pipes were all mined to depletion by 1987 and the property was environmentally reclaimed.

Warm Springs District

The Warm Springs District is located east of Hacks Canyon District on the Kaibab Plateau including Jacob Lake, north of the Grand Canyon. Copper mineralization in a stratiform breccia zone in limey sandstone interbedded in the Kaibab Limestone was discovered in 1885 and several small mines operated into the early 1900s. A small settlement called Ryan became a copper smelting center between 1902 and 1907. In 1909 uranium specimens were found on the Kaibab Plateau. In 1939 some of the copper mines reopened and operated during World War II when more than 14,000 tons of five to seven percent copper ore was shipped to smelters from two properties.

Bentley District

The Bentley District is located north of the Colorado River in the "Arizona Strip" (the strip of land between the Colorado River and the Arizona-Utah state line) on the western part of the Colorado Plateau along the Grand Wash Cliffs.

Copper and minor uranium mineralization occurred in or adjacent to breccia pipes.

About 1853 Mormon settlers were shown silver and copper outcropping by Indians, and they purchased the prospect for a horse and several sacks of flour. Named the Grand Gulch Mine, over the next two decades sporadic shipments of copper and silver ore were made to St. George, Utah.

About 1870 a crude adobe furnace was built with poor results. In 1880 another furnace was built and operated successfully for a few years.

Between 1906 and 1917 an average of 120 tons of concentrates greater than 14 percent copper were shipped monthly to a smelter in Salt Lake City. In 1911 the Grand Gulch Mine was called the "Richest Copper Mine in Arizona."

The mine burned in 1917, having produced 15,701 tons of ore containing 24,349 ounces of silver, 6,651 pounds of copper and 715 pounds of lead. During the same period the Savanie Mine along the edge of another breccia pipe produced some $300,000 in copper.

In the 1970s and 1980s the properties were evaluated for uranium but no economic uranium ore bodies were found.

Mount Trumball District

The Mount Trumball District is located north of the Colorado River and east of the Bentley District. Copper with minor silver, lead, zinc, gold and uranium mineralization was found in breccia pipes and copper in stratiform breccia.

Copper in breccia pipes in the Esplanade Sandstone was discovered in 1875, and the Copper Mountain Mine was worked sporadically until the early 1950s. The mine dumps were reworked between 1955 and 1961. The area was evaluated for uranium in the 1970s, but no economic deposits were found.

Lee's Ferry District

Fine placer gold derived from shale was discovered in Glen Canyon in the Colorado River upstream from Lee's Ferry in 1897. A hydraulic hose system was in operation in 1910, and a steamboat dredge, the Charles H. Spencer, operated in 1912 . The placer operation ended in 1913.

"A Ghost from the Past," one of the last remaining buildings in the mining camp of Mineral Park, this adobe cabin, covered in part with tin and wood, remains as a testimony to what once was. Photo by Ann Ettinger.

THE RAILROAD:
THE ATLANTIC & PACIFIC
AND LATER,
THE SANTA FE

In 1880 Arizona had a population of about 40,000, of which 90 percent resided in the southern part of the Territory.

In the north central part of the Territory there were several small Mormon colonies, and Beale's Wagon Road was the east-west travel route. The area that would become Flagstaff, settled in 1876 in the high timber country, had only a few cabins.

In the west there were the mining camps of Hackberry, Mineral Park, Cerbat and Signal. Supplies came by Colorado River steamers to Hardyville and Aubrey Landing and were transported to the mines by teamsters.

The railroad would open northern Arizona to development of a wide extent of country through which it passed. The railroad settlements grew into towns and small cities which provided services to the surrounding mining districts, ranches and farms. The route was open all year and would shorten the time and distance across the continent and reduce the cost of travel.

The history of the railroad in northern Arizona is one of conquering a wilderness, power struggles with competing railroad moguls, bankruptcies, mergers, difficult economic conditions, fighting the might of the Colorado River and frequent heavy rain, flooding and washouts.

THE ATLANTIC & PACIFIC RAILROAD

In the post-Civil War era, the western United States was ripe for expansion and new railroads were the great civilizers of the 19th Century. Except for stagecoaches and horse and wagon trails, northern Arizona was an isolated wilderness in need of railroad accessibility.

In 1853 Congress commissioned Captain Amiel Weeks Whipple of the Army Topographic Corps to conduct a survey for a proposed transcontinental railroad. Congress eventually decided against the railroad, and subsidized a network of wagon roads intended to improve military and civilian communication throughout the western frontier.

On July 27, 1866 an Act of Congress chartered construction of the original Atlantic & Pacific Railroad (A&P). The vision was to construct a railroad from Springfield, Missouri to the Pacific Ocean. The A&P's first task was how to organize such a vast project over thousands of miles of wild, uncharted Indian country.

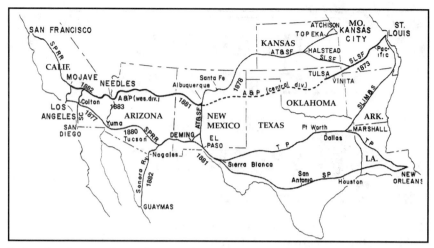

Early railroads across the southwestern United States.   Modified from Myrick, 1963.

The Central Pacific Railroad laid track eastward from Sacramento, California and the Union Pacific Railroad laid track westward from Omaha, Nebraska. They joined on May 10, 1869 with the insertion of the Golden Spike at Promontory, Utah. The Central Pacific owners, the "Big Four:" Huntington, Hopkins, Crocker and Stanford, became very powerful and possessive men.

In 1871 the A&P surveyed a route which roughly followed the 35th Parallel. The route would later become a part of the Santa Fe Railroad's main line connecting Chicago and southern California.

Some 300 miles of rail were laid as far as St. Louis, Missouri but troubled times followed. The financial panic of 1873 forced the A&P into bankruptcy and reorganization yielded a new company, the St. Louis & San Francisco Railway Company (Frisco), which also had cash problems. Principals with the Atchison, Topeka & Santa Fe Railroad (Santa Fe) approached the Frisco offering financial backing with the objective of using the A&P route from Albuquerque to reach the Pacific Coast.

On January 31, 1880 the Santa Fe and Frisco companies signed two agreements. The first split the stock of the A&P in half, giving both the Santa Fe and Frisco equal and joint control of the Santa Fe road rights and privileges. The second agreement between the Santa Fe, the Frisco and their joint subsidiary, the A&P, provided for the construction and finance of the A&P's chartered road from a point near Albuquerque, New Mexico west to the Pacific Coast.

In 1880 Lewis Kingman surveyed a primary route line between the Colorado River and Flagstaff which roughly followed the 1871 survey. Actual work began at the A&P junction near Albuquerque late in the spring of 1880. There were problems with Indians, rough characters and winter weather, but the graders and track layers pushed westward and by the end of 1881 about 190 miles of rail had been laid across western New Mexico and into the Arizona Territory.

The next 360 miles of rail to reach the Colorado River near Needles, California would take three years to build. Lewis Kingman built a head-

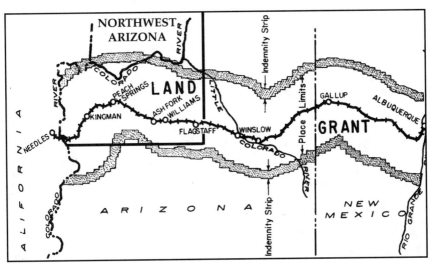

Atlantic and Pacific Railroad land grant.　　　　Modified from Bryant, 1974.

quarters office on a ranch near Bill Williams Mountain and occupied it for more than a year as the railroad was constructed westward.

The Central Pacific became the Southern Pacific (SP) and in the early 1880s pushed its rail lines southward through Los Angeles and eastward through Yuma, Arizona to El Paso, Texas and then eastward another 90 miles where it joined with the Texas & Pacific Railroad (see map, page 82). The SP now controlled the northern and southern overland routes to California and did not want any other railroads diluting its market. The SP objective was to stop the A&P at the California border.

In early 1882 the SP began construction of a rail line from Mojave, California eastward to Needles. The A&P tracks were still 40 miles east of Needles and the Colorado River would need to be bridged. In May 1883 the A&P finally reached the east bank of the Colorado River.

Kingman originally located two routes two miles apart to cross the Colorado River and the upper route was eventually chosen. The building of the Colorado River bridge was a major engineering effort.

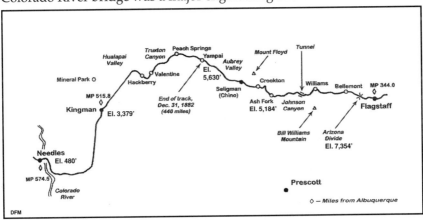

Atlantic and Pacific Railroad, Needles to Flagstaff.　　　　Modified from Myrick, 1998.

Laying of the Atlantic & Pacific tracks, circa 1882.
Courtesy of the Mohave Museum of History and Arts, Kingman, Arizona

In 1883 there were no dams to control the flow of water along the entire length of the Colorado River. Spring runoff from the Rockies swelled the river with treacherous, swirling currents, carrying with them tree stumps and other debris. Along with the rising waters were areas of quicksand from which bridge pilings were swept away as soon as they were set in place.

There were thoughts of postponing the Colorado River bridge work until the waters had receded but work continued until, by the end of June, only a 20-foot gap remained to complete the bridge. Then a surge of high water swept away 280 feet of pile trestle. Work continued to rebuild the bridge.

Meanwhile, A&P track crews began laying track between Needles and the west bank of the Colorado River where it met the bridge. At the same time, SP crews were constructing a 14-stall roundhouse, a hotel, and a magnificent depot that could accommodate 500 people in Needles.

In early August of 1883 the bridge was completed and on August 9th the "35th Parallel Trans-Continental Line" was declared officially open.

Williams and Peach Springs were division points along the line, each with a roundhouse and repair facilities.

LEWIS KINGMAN

Lewis Kingman was a civil engineer who became a prominent railroad builder in the southwest and Mexico. He helped build part of the Atlantic & Pacific, now the Santa Fe railroad, between Albuquerque, New Mexico and Mojave, California, a distance of 915 miles. He built 1,353 miles of other SF system lines. As engineer for the Mexican Central, he constructed nearly 1,500 miles of railroad in Mexico. The town of Kingman, Arizona was named after him.

Born in Massachusetts in 1845, Lewis Kingman was raised on a small farm. At age 17 his father arranged for him to study and work for a civil engineering firm in Boston, paying $100 a year for the boy's instruction. This is where Lewis got his first railroad construction experience.

In 1864 Lewis, 19, went to Pennsylvania during the oil boom and learned to drill oil wells, investing in this new business of petroleum. When the oil boom went bust in 1868 he was hired on by the A&P Railroad in St. Louis and started as a transition man on a survey crew. Within a few weeks he was given 12 miles of railroad to build.

In 1870 Kingman did the initial survey of the A&P route to Albuquerque. In 1871 he conducted railroad route surveys across little-explored New Mexico and Colorado, which would take the next two years. At the same time other surveys were made across northern Arizona.

Between 1873 and 1876 Kingman contracted to survey for the government. In 1876 he concluded his government contract and went into the mercantile business in Santa Fe, New Mexico. His business failed in 1877, and he was then hired by the Santa Fe to survey new routes in New Mexico and Colorado.

In 1879 he resurveyed the 1870 route from Colorado to Albuquerque.

In early 1880 Kingman was instructed by Santa Fe's chief engineer in Santa Fe, New Mexico to locate and survey a primary line along the A&P right-of-way following the 35th Parallel as quickly as possible following the earlier 1871 survey. He reviewed the earlier survey and accepted some locations and revised others.

His orders were to travel from Flagstaff to the Colorado River and to survey a route from near Fort Mohave eastward where he would join up with another survey near Flagstaff. Kingman's crew consisted of 20 men and a few mules.

Kingman located two survey lines to cross the Colorado River near Needles and proceeded eastward.

The survey was through blistering hot desert. Water, hauled in barrels, was transported at night from the Colorado River by a four-mule team, a longer trip as each new camp was made. For the first 46 miles, the rule was one pint of water for each man for washing each morning, and no more than was absolutely necessary for drinking and cooking. After that, Kingman tried to locate his camps near springs and he avoided sickness by boiling the water. Kingman described camp life while on the survey:

> I had good mule teams and a saddle horse, was well supplied with provisions; had tents, a cookstove and water barrels. I kept a good cook and served good meals. My rule was to have breakfast at sunrise, reach the end of the line as soon as we could (which end of the line would, of course, be some difference from the camp), work until 12 o'clock, give the boys an hour for dinner, usually make hot coffee and have a good basket of lunch and some canned goods for relish. Then we worked in the afternoon until 5 or 5:30, quitting in time to reach camp just before sundown. The boys always enjoyed eating supper just before dark, after which we assembled around a good campfire. In a crowd of fifteen men there are always good story tellers and one or two with a little more wit than the others, and usually a majority who are inclined to

enjoy life and make the others comfortable. Then there is always one, sometimes two, who think themselves most miserable and are inclined to find fault with everybody and everything. These always get what they deserve and find that the world is just what they make it.

In selecting the route Kingman and a mountain man guide took their burros and traveled in the field for five days at a time.

During the survey there was rain, hail and lightning. One incident of note was when Hualapai Indians stole three of Kingman's burros, which he bought back for five dollars.

Later Kingman wrote about the survey:

In all my hard work in Arizona, I enjoyed every detail; it was hard and strenuous, but I liked it, and the solitude of that large country was no hardship to me. It was a part of creation and I had no reason to think that it was anything but good. I tried to adapt myself to circumstances surrounding me and make the best of them. Looking back, the four years spent on this line was full of interest. It was a part of my education and experience in locating, construction and handling men and contractors and building of a railway.

The survey party reached Flagstaff about November 1 and shortly thereafter construction contracts were signed for building the railroad west of Winslow, Arizona.

Kingman superintended the building of the railroad from Winslow, Arizona to Beale Springs, opposite the present town of Kingman.

Kingman's headquarters office was on a ranch near Bill Williams Mountain, which he occupied for more than a year as the tracks were laid westward. Huge quantities of construction materials were stockpiled to build the 600-mile segment, including 800 carloads of rail.

By February 1881 100 miles of line were completed. Minor problems were encountered with the Indians, who tore up portions of the track at night.

By September Kingman's crews had laid 236 miles of track. Grading crews initially were mostly Irish with some Mexicans. The A&P paid graders and tracklayers $2.25 per day; spikers and iron layers received $2.50 per day.

Kingman was appointed chief engineer of A&P in 1882 and held that position until April 1, 1883 when he resigned and accepted a position as chief engineer for the Mexican Central, building 452 miles of railroad in northern Mexico in little over a year.

About July 1, 1884 Kingman was rehired by the Santa Fe in various positions at $350 per month. He was let go at the end of 1888 because of bad railroad economic conditions in the west.

On May 1, 1895 Kingman was appointed chief engineer of the Mexican Central Railroad, headquartered in Mexico City. Between 1895 and 1907 Kingman built 994 miles of railroad in Mexico, when the Mexican Central was taken over by the Mexican Government and became a part of the National Railways of Mexico. Kingman continued to work in various jobs for the National Railways of Mexico until his death in Mexico City on January 23, 1912 from pneumonia at the age of 66.

THE FIRST TRAIN

The new business of railroads in the west brought fierce competition and power struggles between the Atlantic & Pacific and the Southern Pacific. The war of attrition between the two railroad giants can be exemplified as follows:

When the first passenger train arrived at Needles from San Francisco on October 22, 1883, the SP halted its coaches at the station and the passengers had to transfer their own belongings across a long platform to waiting A&P cars. Only the sleeping cars were interchanged.

Both the A&P and SP developed their own local business, neither of which was enough to pay expenses.

By the close of 1883 it was apparent that, due to the SP strategy of isolating the A&P route at Needles, in its present condition the A&P line was virtually worthless.

Both the A&P and the SP had a lot to lose, so the Santa Fe (A&P's parent company) and SP sat down and reached an agreement that became effective October 1, 1884 where the A&P leased SP's route from Needles to Mojave, California, including trackage rights from Mojave to Oakland and San Francisco.

In 1884 the Santa Fe was involved in the construction of a rail line from San Bernardino to Barstow. This was completed in November 1885 which gave the A&P direct access to the port of San Diego. This was followed by rail construction to Los Angeles which was completed in 1887.

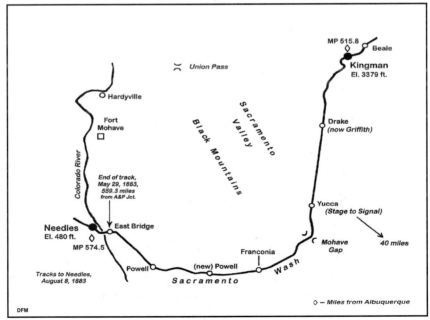

Atlantic and Pacific Railroad, Needles to Kingman.   Modified from Myrick, 1998.

Spring flooding of the Colorado River was a continual problem for the A&P. The bridge built 2.5 miles below Needles washed out in May 1884 and, for the next two months, all passengers and baggage had to be ferried across the river on flatboats until a new bridge was built.

Bridges were again swept away in 1886 and 1888, during which time the A&P had to maintain standby river boat service during periods of high water. The A&P decided to construct a new single-track cantilever bridge downstream at Red Rock. Work began sinking piles September 1888 and, with delays due to summer heat, actual construction of the bridge began on February 4, 1890 and was completed in April. New approaches to the bridges required 13 miles of railroad tracks from Needles to a point beyond Topock, Arizona. Two hours after the new route was opened on May 10, 1890 a flash flood washed out two miles of rail. The final bridge test was on June 14 when ten large locomotives were run out on the bridge and the framework held up under the weight.

The steam locomotives of this time used coal to heat their boilers and water sources along the line were necessary. On the Colorado Plateau there is little subsurface water. The A&P looked for springs along its route and also built reservoirs when possible to catch rain and melting snow. The water was then piped to a rail siding for use in the locomotives.

Pipelines from reservoirs and springs delivered water to Flagstaff, Williams, Peach Springs Station and Kingman. Tank cars from Williams supplied water to Ash Fork and Seligman and each day five cars of water from Mellen or Needles arrived at Yucca. This added $90,000 to the annual operating costs.

Management decided that longer engine runs were acceptable and Seligman was designated the new division point in October 1897, which eliminated both Williams and Peach Springs.

A combination of high maintenance and competition from the Southern Pacific and Union Pacific to the north, the Southern Pacific to the south, and the Depression of 1893 saw the Santa Fe and Frisco, followed by the A&P, go into the hands of receivers. The reorganization gave the Santa Fe the entire operating Western Division of the A&P, which included the route from Albuquerque, New Mexico to Needles, California, and the leased line of the SP from Needles to Mojave, California.

THE SANTA FE RAILROAD

On June 30, 1897 the A&P ceased to exist. An Act of Congress, titled the Santa Fe Railroad, conveyed all A&P property to a new Santa Fe subsidiary and operated as such for the next five years.

On January 16, 1898 rail service was disrupted when a fire started in the Fairview Tunnel west of Williams. Passenger trains were diverted over other lines and many freight trains were held up. The fire was fully extinguished, repairs made, and service resumed on February 8.

The Santa Fe, in addition to freight trains, operated a twice-weekly passenger train between Chicago and California during the winter, making the trip in 71 hours.

Plans were made to replace all A&P rails and develop new water reservoirs which included Seligman and a steel dam at Ash Fork.

Intermittent droughts caused problems. Engines would stand idle for days because there was no water. In September 1898, water was in short supply in Flagstaff and freight engines had to be detached from their westbound trains and travel to Bellemont for water before returning to resume their normal journey.

Along with the droughts came periods of localized heavy rains which washed out short sections of track.

At the end of 1900 the Santa Fe ran three pairs of daily passenger trains between Chicago and Los Angeles four times each week with fares as low as $30.

In January 1901 coal miners at Gallup, New Mexico mines went on strike which spurred the Santa Fe to make the decision to use fuel oil to fire its locomotive boilers. By the fall of 1901 Santa Fe oil-fired locomotives operated eastward from Needles to Seligman. It wouldn't be long before fuel oil was used on the runs east of Seligman.

In 1901 a spur line was built from Williams to Grand Canyon Village.

On July 1, 1902 the title and operations of the former A&P system were transferred to the parent Atchison, Topeka & the Santa Fe, and the route formed part of the Santa Fe's "grand division," called the Coast Lines. The passenger train was called "The California Limited."

The newer locomotives became larger and heavier and it became necessary to strengthen the Colorado River bridge. The bridge was strengthened in 1901, 1910, and 1931-37.

With the Santa Fe takeover of the A&P line across northwest Arizona came the responsibility of upgrading to accommodate growing traffic volume and the yearly storm damage along the line.

In July 1904 a heavy storm struck Truxton Canyon causing heavy damage. The roadbed was relocated along higher ground to avoid future washouts.

On July 9, 1905 Death Valley Scotty hired a special train, "The Coyote Special," for $5,500 to take him from Los Angeles to Chicago in 46 hours. The trip was made in 44 hours, 44 minutes.

In February 1905 a rainstorm washed out a quarter mile of rail and pumphouse in Railroad Canyon west of Kingman.

During the same period worn rails from the A&P era were being replaced. In 1905 a number of derailments occurred when old rails broke.

The maximum eastbound and westbound grades were 1.42% and helper engines were needed to pull the grades. It was slow going.

The single daily transcontinental train was increased to three pairs of trains in March 1907. As the population in southern California grew, freight traffic in both directions increased, especially eastbound perishable tonnage.

The decision was made to construct a double track across northern Arizona along with realignment of much of the original main line, requiring extensive grading to reduce curvature and gradients.

The Santa Fe budgeted $4 million for roadway improvements in northern Arizona. Construction of the double tracks in northwest Arizona be-

A Santa Fe Eating House, Kingman, Arizona, operated by the Fred Harvey Company. The approaching train, traveling eastward, has just climbed out of Kingman Canyon and is preparing to stop at the passenger depot behind the photographer, circa 1909.
Courtesy of the Mohave Museum of History and Arts, Kingman, Arizona.

gan when a camp was established in Flagstaff in March 1910. Most of the available houses were rented by engineers and office workers. Most line workers resided in Williams.

Steamshovels and rock gangs removed thousands of cubic yards of rock from cuts and transported some of it for fill material elsewhere along the new line.

There were problems. Several fires were started from locomotive sparks. Premature dynamite explosions killed a number of men. Runaway cars caused some serious injuries and deaths.

By the end of 1911 double tracks were in use between Flagstaff and Williams.

The second track was built between Ash Fork and Seligman in 1912 and 1913, and by 1914 reached Yampau Summit (El. 5,630 feet). Work was suspended until 1922 when contracts were signed for second track construction in Mohave County through Kingman.

On August 14, 1923 the Santa Fe had 459 miles of double track running between Winslow, Arizona and Barstow, California.

During a drought in the fall of 1934 the Santa Fe delivered 14 tank cars of water to Flagstaff daily for several months. Heavy winter snows melted and filled the city reservoirs in 1935.

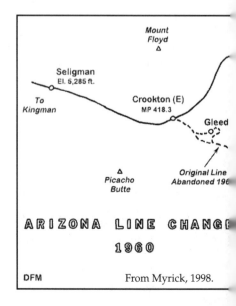

ARIZONA LINE CHANGE
1960

DFM   From Myrick, 1998.

90

Santa Fe locomotive No. 3759, a type 4-8-4, built by Baldwin in 1927 is now located in Locomotive Park across from the Power House in Kingman, Arizona.   Photo by L.J. Ettinger.

Preparing for war, a new double track and heavy duty bridges were planned in 1941. Construction started in September 1942 and was completed in 1945.

In 1960 Santa Fe relocated 44 miles of line west of the Arizona divide from Williams to Crookton, eliminating the slowest part of its overland line by reducing grades and curves. The cost was $22 million and took 17 months to complete. One cut is two miles in length, reaching a depth of 115 feet.

PASSENGER TRAVEL

In the early 1900s Santa Fe advertising was directed toward tourist passenger traffic. The route traversed Indian country, the Continental Divide, the Painted Desert, Canyon Diablo Bridge and side trips to nearby national parks. Harvey Eating Houses were constructed to serve passengers. Trains were named "The Missionary" (1915), "The Navajo" (1915), "The Scout" (1916), "The Chief" (1926), "Grand Canyon Limited" (1929), and "The Hopi"

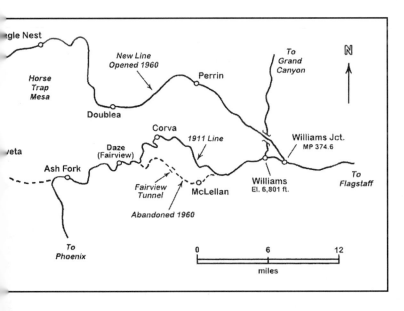

Santa Fe

SANTA FE
ALL THE WAY

TIME SCHEDULES
OF PRINCIPAL TRAINS

The Atchison, Topeka and Santa Fe Railway System

From Myrick, 1998.

(1929). In 1936 a new blunt-nosed diesel-powered train called "The Super Chief" began a 40-hour trip between Los Angeles and Chicago. The "El Capitan" followed in the same year.

Since 1960 passenger traffic dwindled with the rise of commercial air travel. In 1964 Santa Fe was running three passenger trains, "The Chief," "The Grand Canyon," and "The Super Chief-El Capitan," over its northern Arizona route.

In 1969 the Grand Canyon spur line was eliminated.

Amtrak took over passenger service from Santa Fe on May 1, 1971 and only one train operated between Chicago and Los Angeles over the Santa Fe's Arizona line, using the names "El Capitan" and "Super Chief." In May of 1974 the name of the train was changed to "Southwest Limited," and renamed "Southwest Chief" in 1985 when new superliner cars were installed.

The Santa Fe and Burlington Northern railroads merged in 1995 with the BN as the surviving company. Today most of the rail traffic are freight trains. Passenger service is provided by Amtrak.

THE FRED HARVEY COMPANY

Fred Harvey was born in London, England and migrated to New York City at the age of 15. He began working in restaurants during the Civil War and then for the railroad following the war. The trains lacked refrigeration and dining cars and, with both restaurant and railroad experience, he saw an opportunity.

In 1876 he founded the Fred Harvey Company. Employing young and attractive women, he opened his first railway restaurant in Topeka, Kansas. Well-prepared meals, spotless dining rooms and courteous service would become the Harvey House trademark.

Harvey would build a restaurant every 100 miles along the Santa Fe route. Over the next 10 years 16 Harvey Houses were opened, including restaurants, hotels, and lunchrooms near Santa Fe stations in Kansas, Oklahoma, Texas, New Mexico, Arizona and California.

The Harvey Girls were one of the Harvey House trademarks. Young women from good homes in the East were recruited for their intelligence, culture and refinement. His advertisements in Eastern papers stated, "Wanted: Young Women 18 to 30 years of age, of good character, attractive and intelligent, as waitresses in Harvey Eating Houses in the West. Good wages, with room and meals furnished."

Their uniforms were crisp white aprons and bows over tailored high-necked black shirtwaists. They were housed in dormitories overseen by strict housemothers. They welcomed and coddled travelers.

The Harvey hotels were trackside palaces in the wilderness and the Harvey Girls earned $17.50 a week, serving as many as six trainloads of diners a day.

Harvey Eating House and Harvey Girls, Kingman, Arizona, circa 1900.
Courtesy of the Mohave Museum of History and Arts, Kingman, Arizona.

In the restaurants passengers would find tables set with Irish linen and food comparable to that in the best Eastern hotels.

Thousands of Harvey Girls married ranchers, railroad men and cowboys and remained in the West. Many of their sons were named "Fred" or "Harvey."

In the 1890s the Santa Fe began including dining cars on some of its trains and Fred Harvey was awarded the contract to serve food in these cars.

Harvey meals included as many as seven entrees, with seconds, for 75 cents. Prices were raised to $1.00 in 1920. Menus were coordinated to avoid duplication on a trip.

Fred Harvey died in 1901. The press dubbed him "Civilizer of the West," and an article from the 1880s said he "made the desert blossom with beefsteak and pretty girls." Will Rogers said of him, "He kept the West in food and wives." His sons took over the company through the 1930s.

In 1902 the Fred Harvey Company hired Mary Elizabeth Jane Colter as company architect to design and decorate the Harvey Houses (see the Grand Canyon Chapter, page 120). Her Pueblo and Mission styles of architecture became famous.

In 1904 the Fred Harvey Company joined with the Santa Fe in planning to develop the Grand Canyon Village on the South Rim.

In 1906 the Fred Harvey Company was incorporated, and by 1917 there were 100 Harvey Houses.

The Fred Harvey Company became the principal concessionaire at the Grand Canyon's South Rim in 1920, and began marketing Native American arts and crafts.

After World War I, with better roads and more automobiles, travelers could spend more time touring. The Harvey Company began packaging motor trips to Native American villages and the Grand Canyon.

Business slowly declined through the 1920s due to increased use of the automobile and air travel. During the Great Depression of the 1930s fewer and fewer people traveled for any reason.

During World War II Harvey House business resurged as troop trains needed meals for the soldiers.

In 1945 Metro-Goldwyn-Mayer released the movie "The Harvey Girls," starring Judy Garland. The sound track included the Oscar-winning song "On the Atchison, Topeka and the Santa Fe."

By the 1950s, to be competitive, the Santa Fe reduced passenger service in favor of freight trains, and the Harvey Company began to focus more on resorts and national parks.

In 1968 the Hawaii-based Amfac Corporation bought the Harvey Company and the Fred Harvey Company ceased to exist. In 2002 Amfac Parks and Resorts was renamed Xanterra Parks and Resorts (a Colorado-based company), and continues today as the Grand Canyon concessionaire.

MAN
AND THE
GRAND CANYON

The Grand Canyon, an awesome array of precipices, amphitheaters, buttes, slopes, spires, temples and shapes carved by the enormous power of the Colorado River, is 277 miles long, a mile deep and as much as 18 miles wide. In the canyon one can see red, gold, pink, green, rust, orange, mauve and many other colors depending on the time of day.

Man's relationship with the Grand Canyon spans more than 4,000 years from the first view of the canyon by Native Americans, followed by the Spanish explorers of the mid-1500s, and then the trappers and traders of the early 1800s. In the mid-1800s Jacob Hamblin and Mormon settlers came to the Grand Canyon country followed by United States Army surveyors. Major John Wesley Powell's expeditions down the Colorado River began in 1869, followed by Lieutenant George Wheeler's survey in 1877 and Clarence Dutton's survey in 1880.

Between 1883 and 1900 William Wallace Bass, John Hance, Louis Boucher, Ralph Cameron, Pete Berry and others arrived at the Grand Canyon as prospectors. Partnerships were formed and they improved old animal and Indian trails and located mining claims over copper, gold and asbestos deposits within the canyon. They quickly realized there was more money to be made with tourism than with the hardships associated with mining. Bass, Hance, Boucher, Cameron and Berry each occupied a part of the south rim, each with a trail into the canyon.

In 1901 the Santa Fe spur line arrived at the South Rim along with the Fred Harvey Company whose goal was to completely control the development of all tourism business.

The Kolb Brothers, Emery and Ellsworth, arrived at the South Rim in 1902 and set up their photography studio in 1903. Then began the classic battle between big business and the "little guy."

Hance, Boucher, Cameron, Berry and the Kolb Brothers were all "survivors," and all played a significant role in the early development of the canyon.

Today little remains of their camps, but their toil and trails remain a legacy to their presence.

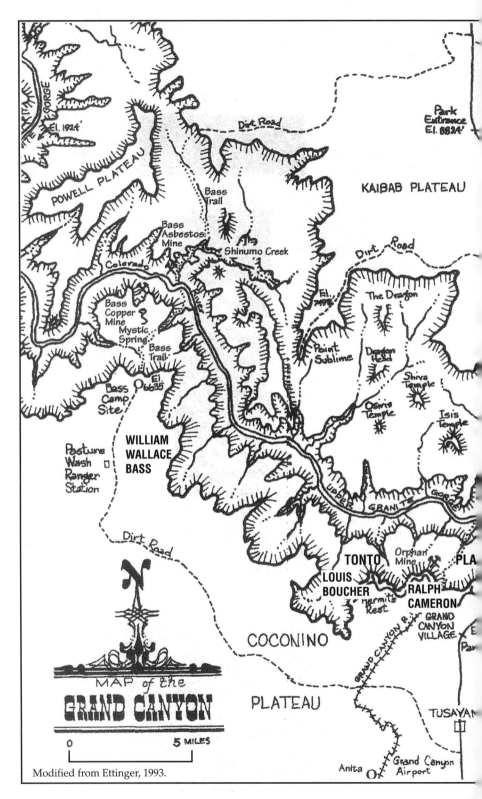

WILLIAM
WALLACE
BASS

MAP of the
GRAND CANYON

0   5 MILES

Modified from Ettinger, 1993.

TO JACOB LAKE
& KANAB

Dirt Road

Navajo
Indian
Reservation

PAINTED
DESERT

Point
Imperial

Grand
Canyon
Lodge

Kwagunt
Butte

NORTH
RIM

El.
8390'

Chuar
Butte

Little Colorado River

El. 2720

WALHALLA
PLATEAU

Salt
Mine

Buddha
Temple

Dana
Temple

Brahma
Temple

Juno
Temple

Zoroaster
Temple

El Tovar
Cape Royal

Jupiter
Temple

Lava
Butte

Phantom
Ranch

Venus
Temple

Apollo
Temple

RIM

Wotan's
Throne

Vishnu
Temple

Hance
Asbestos
Mine

Cedar M
El.
7053

Last
Chance
Mine

DESERT
VIEW
El. 7498'

Horseshoe
Mesa

Grandview
Point

JOHN
HANCE

Hance
Ranch

Park
Entrance

Entrance

SOUTH
RIM

Tusayan
Museum
& Ruins

PETE
BERRY

..... TRAIL
- - - DIRT ROAD
PAVED ROAD
++++ RAILROAD

TO WILLIAMS,
FLAGSTAFF
& I-40

TO CAMERON
& U.S. 89

EARLY VISITORS TO THE GRAND CANYON

   Archaeological evidence of the Desert Archaic Culture suggests that pre-historic North American Indians were the first to descend into the depths of the Grand Canyon about 4,000 years ago. They were nomadic hunter-gatherers living in small family groups and they left evidence in caves in the Redwall Limestone, including small animal fetishes made from willow twigs. The Anasazi Indians (ancestral Puebloan Basket Makers) moved into the Grand Canyon area for a period of about 700 years between 500 to 1200 A.D. They were farmers and also hunted deer, rabbit and bighorn sheep. They were able to live a more settled life, occupying both the north and south rims as well as areas within the canyon. The Anasazi abandoned the Grand Canyon country by 1300 A.D., probably as a result of the Great Drought of 1276 to 1299.

"Vishnu's Temple" by W.H. Holmes. From U.S.G.S. Monograph 2, 1882.

In the 1300s the Pai, a hunter-gatherer tribe and forerunner of the present-day Hualapai and Havasupai tribes, moved into the Grand Canyon area. At the same time Southern Paiute roamed the north rim.

In Spain there was a legend of a Bishop who fled from Arabs, taking with him vast treasures and establishing seven cities in the unexplored world.

Spaniards in Mexico in the 1530s heard stories of golden treasures to be found in the Seven Cities of Cibola, somewhere far to the north. Believing the stories to be true, in 1540 Francisco Vásquez de Coronado took an expedition of 300 well-armed young Spaniards with horses and cattle in search of these riches.

Finding only Indian pueblos in what would become the American southwest, Coronado heard Indian reports of a large river lying to the west. He sent Captain García López de Cárdenas and 25 men in search of it.

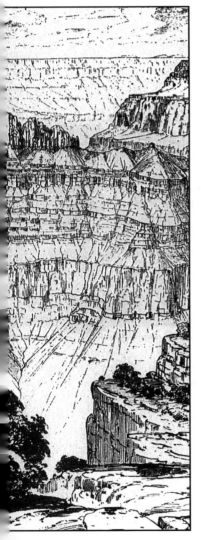

After 20 days, guided by Hopi Indians, Cárdenas arrived at the south rim of the Grand Canyon, probably between what is now Moran Point and Desert View, becoming the first European to view the canyon.

The party spent a week trying to find a crossing down into the canyon without success, and then retreated to join up with the main Coronado expedition to the east.

The Grand Canyon would not be seen again by European explorers for another 236 years when, on June 26, 1776, Fray Francisco Tomas Garcés, a Franciscan missionary, was guided by Havasupai Indians to the south rim to view the canyon. Garcés was the first man to refer to the river as "Colorado" which translates from Spanish to "red-colored" because of the brick-like hue of its silt-laden water.

Navajo hunter-gatherers reached the Grand Canyon area sometime between A.D. 1700 and 1800. Over 2,000 archaeological sites have been found, many with pictographs and cave paintings along with rock granaries.

THE 1800s

In 1821 Mexico gained independence from Spain, obtaining sovereignty over the region which included the Grand Canyon. Mexico allowed access to this country to traders and fur trappers who were the first Americans to see the Grand Canyon sometime between 1826 and 1828.

The war between Mexico and the United States began in 1846 and ended with the Treaty of Guadalupe Hidalgo on February 2, 1848 whereby Mexico ceded a large tract of its northern territory to the United States. This included the entire Grand Canyon country.

Subsequently, the U.S. Government made plans to send Army surveyors to chart the unknown and almost inaccessible territory and, in the spring of 1858, a military exploration expedition headed by Army Lieutenant Joseph Christmas Ives viewed the Grand Canyon for the first time. His adventures were published in 1861 as "Report Upon the Colorado River of the West." In the 1850s the canyon was referred to as "Big Cañon" and "Grand Cañon of the Colorado." By 1865 "Grand Canyon" was widely used.

Dispatched by Brigham Young, Mormon missionary Jacob Hamblin explored the Grand Canyon region to determine future settlement potential. He established peaceful relations with the Navajo and Hopi Indians between 1858 and 1864. He found Colorado River crossings at what would later become Lee's Ferry to the east and Grand Wash Cliffs at the western lower end of the Grand Canyon. He descended into the canyon and visited the Havasupai, Hopi, Paiute, Shivwits and other Indians of the surrounding plateaus. He found that there was no easy way to cross the Colorado River within the canyon.

Government surveys of the Grand Canyon area did not begin until after the Civil War. Major John Wesley Powell led a series of expeditions down the Colorado River to study the Grand Canyon geology beginning in 1869.

In May of 1869 Powell, poorly funded and with no illustrator or photographer, left Green River Station, Wyoming. One of the expedition's four boats carrying most of their food supplies capsized, forcing the party to proceed with little time to collect necessary data or to plot the course of the river. During the 100 days on the river the survey party did learn about survival on the rough waters of the Colorado River.

In 1872 Army Lieutenant George Montague Wheeler led an expedition upriver along the Colorado 52 miles into the western part of the Grand Canyon. His team took the first photographs of the canyon.

Powell's second trip down the Colorado into the Grand Canyon left Lee's Ferry on August 17, 1872 and ended at Kanab Creek, about halfway through the canyon. His boats were redesigned and the party resupplied at several points along the way. Powell also brought a photographer and an artist to show the world the beautiful landscapes he encountered on his trip down the canyon. Between 1871 and 1873, over 20,000 photographs of the Grand Canyon were shot from the Colorado River, side canyons and the rims.

There were no digital cameras in those days. Nearly one ton of equipment including darkroom apparatus and chemicals, as well as glass plates and large cameras, had to be moved into position for each shot. The equipment was carried in boats or on mules to each location. When a spot was found where three or four views could be photographed the equipment was unpacked, a tent used as a darkroom was erected, and the photographs taken and processed. Everything was then repacked and new photograph sites were sought. Problems included gusts of wind blowing dust and sand on the photographic plates.

Powell's expeditions brought him national fame. He used photographs to illustrate his lecture tours and derived funding for his projects by selling rights to reproduce some 650 of the 1,400 stereographs taken during his expeditions.

Clarence Edward Dutton, of the U.S. Geological Survey, conducted a study of the Grand Canyon in 1880 and 1881 and published the first geological monograph of the area, "Tertiary History of the Grand Cañon District with Atlas" (1882). William Henry Holmes, an artist, accompanied the survey and two of his panoramic line drawings are shown in this chapter.

The early 1880s brought prospectors to the Grand Canyon region and led to much of its early development. Minerals discovered included gold, silver, lead, copper and asbestos, but deposits were small and most were uneconomic.

The Atlantic and Pacific Railroad arrived at Flagstaff in 1882 and with

rail accessibility there was interest in bringing tourists to visit the Grand Canyon. The same year Senator Benjamin Harrison introduced the first bill to create Grand Canyon National Park, but without success.

John Hance was probably the Grand Canyon's first white settler in 1883, claiming land on the south rim where he built the first tourist facility, a log cabin, and advertised guided tours into the canyon. Stagecoaches began bringing tourists to the Grand Canyon in 1883.

Ralph Cameron arrived at the Grand Canyon in 1883 and over a period of years located mining claims covering 13,000 acres of land in and around the canyon, including most of Indian Gardens and all of Bright Angel Trail, designating it as a toll road. He also improved an old Havasupai Indian Trail, making it safe for travel to the bottom of the gorge. Before the railroad arrived at the south rim he constructed a hotel close to the head of the trail. Cameron became a U.S. Congressman and later U.S. Senator from Arizona. His claim on Bright Angel Trail resulted in numerous lawsuits with the Santa Fe Railroad and the Harvey Company, which lasted until 1928 when the U.S. Government assumed control.

In 1883 William Wallace Bass arrived at the Grand Canyon looking for lost gold but quickly found that the tourist business offered greater rewards.

In 1884 the first hotel, the Farlee, with two bedrooms, opened for business on the south rim of the western Grand Canyon north of Peach Springs. The Farlee operated until 1889.

In subsequent years others constructed hotels and tourist camps and, along with the prospectors and miners, built and upgraded old Indian trails into the canyon. Mules first brought to the area by prospectors were used for trips into the canyon.

In the late 1880s railroad and coal barons hatched a plan to build a Colorado River grade railroad from the Rocky Mountains to the Pacific Ocean through the Grand Canyon whereby coal would be quickly and efficiently hauled to the West Coast, and mineral wealth from California and the Southwest would be hauled eastward to railroads linking Eastern markets. The Denver, Colorado Cañon and Pacific Railroad was established and an en-

"Panorama from Point Sublime — Looking West" by W.H. Holmes.

gineering feasibility study of a railroad following the Colorado River from Denver, through Utah and Arizona to San Diego was approved.

In 1889 Robert Brewster Stanton set out on the third boat passage down the Green and Colorado rivers through the Grand Canyon. In Marble Canyon three of his party drowned, including the president of the railroad company. The survey was completed in 1890 and the route was found to be feasible; however, the capital needed to build the railroad never materialized.

The same year President Benjamin Harrison created the Grand Canyon Forest Preserve. Bright Angel Trail opened in 1891 and tourist mule rides to the bottom of the canyon began.

U.S.G.S. Monograph 2, 1882.

103

The fourth and fifth boat trips through the canyon were in 1896 and 1897, both for trapping and prospecting as well as for the excitement.

John George Verkamp, working for the Babbitt Brother's Trading Company, brought the first curios to the Grand Canyon from Flagstaff in 1898. He rented one of the Bright Angel Hotel tents, but business was so slow he left after a few weeks.

THE 1900s

Before rail service to the south rim, the only transportation to the Grand Canyon was by stagecoach, a rugged 20-hour trip from Flagstaff at a cost of $20. A tent camp at Grandview Point attracted most of the tourists.

The Santa Fe Railroad offered easier tourist access by constructing a 65-mile spur line from Williams, Arizona to the South Rim of the Grand Canyon in 1901. The rail trips took less than three hours and cost $3.95. By design the tracks ended at a place along the rim which could be developed by the Fred Harvey Company and was named Grand Canyon Village. The plan was to put all of the other tourist operations out of business. Passenger service operated until 1968. In 1902 the first auto arrived at the Grand Canyon.

Ellsworth and Emery Kolb established their photography business at the South Rim in 1903. In the same year David Rust provided services for tourists by pitching several tents near present-day Phantom Ranch. He called it "Rust's Camp."

In 1904 the Fred Harvey Company went ahead with plans for a hotel at the South Rim to accommodate the expected influx of tourists.

Architect Charles F. Whittlesey of Chicago designed a huge 80-room, dark-wooden structure with a Swiss Chalet appearance. Named El Tovar after Don Pedro de Tovar, an explorer with Coronado in 1540, the hotel opened in mid-January 1905. Mary Elizabeth Colter's Hopi House was built to serve as the main sales area for Native American arts and crafts. The Hopi House opened a few days before the El Tovar on January 1, 1905.

With the railroad, the automobile and road access came more tourists and the real need to protect the natural state of the Grand Canyon for future generations. First designated as a Forest Reserve in 1893, President Teddy Roosevelt revisited the Grand Canyon in 1903 and declared the canyon a Game Reserve in 1906.

Thrill-seeking boat trips through the canyon followed in 1903, 1907, and 1909. The use of better designed boats greatly reduced the number of capsizings and drownings that occurred in the rapids during the first trips down the river.

Without a water supply at the South Rim it was necessary to deliver water in railroad tankers from Del Rio on a daily basis, a distance of 120 miles.

John Verkamp returned in 1905, built a store east of the Hopi House, and went into business for himself in January 1906. Verkamp operated his business of selling postcards and Native American rugs successfully in spite of efforts by the Sante Fe and Fred Harvey Company to shut him down. John's son, Jack, and daughter took over management of the store in 1936. Today Jack's son, Michael, manages the business in the same building built in 1905.

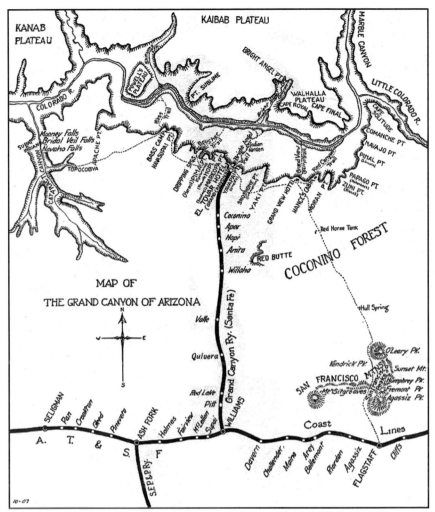

From "Titan of Chasms," a publication of the Santa Fe Passenger Department, 1908.

In 1908 the federal government established the Grand Canyon National Monument and prohibited new mining claims within its boundaries.

Lookout Studio was built by the Fred Harvey Company in 1914 near the Kolb Brother's studio to compete with them. Designed by Mary Elizabeth Colter, it was constructed of rough-cut limestone blocks.

In 1911 the 26th and 27th person to boat the Colorado River through the Grand Canyon were part of the Kolb brothers trip. The resulting photographs, motion picture, a book entitled "Through the Grand Canyon from Wyoming to Mexico," and lectures did much to promote tourism of the Grand Canyon.

In 1915 106,000 tourists visited the Grand Canyon

Grand Canyon National Park was created on February 26, 1919. Changes in the park boundaries in 1927 and 1975 doubled the park size.

Phantom Ranch was built along Bright Angel Creek about one half mile north of the Colorado River in 1922 by the Fred Harvey Company. The ranch was designed by Mary Coulter to provide food, lodging and comfort for tourists who rode mules or hiked to the canyon bottom.

The first detailed survey of the Colorado River through the Grand Canyon was in 1923 by the U.S. Geological Survey.

The North Rim was developed for tourists with the building of the Grand Canyon Lodge in 1928. The Watch Tower at Desert View on the South Rim was built in 1932.

In 1931 the Santa Fe Railroad developed a new water source at Indian Garden Springs, laying a 2.5 mile pipeline and pumping station to the South Rim. It was no longer necessary for the daily train to bring water to Grand Canyon Village.

On March 31, 1933 at the height of the Great Depression, the Civilian Conservation Corps (CCC) was signed into law by President Franklin D. Roosevelt. The first CCC group arrived at the Grand Canyon on May 29, 1933 and four CCC companies of about 200 men each worked on the South and North Rims and in the canyon bottom until 1942. They improved the Bright Angel and other trails, and constructed roads, overlooks, fences, rest houses, the trans-canyon telephone line, rock walls, buildings, picnic shelters, campgrounds and bridges. Men signed up for six months at a salary of $30 a month, $25 of which was sent home to their families.

Men of the Civilian Conservation Corps near the Grand Canyon, 1935.
Courtesy of the Mohave Museum of History and Arts, Kingman, Arizona.

On June 22, 1935 the Bright Angel Lodge opened and offered more economical facilities for tourists. The lodge was expanded later the same year.

Ten-car water trains were sent from Flagstaff every other day for several weeks to meet the added water needs.

The first solo boat trip down the Colorado River through the Grand Canyon was in 1937, and the first commercial river rafting operation began in 1938 and carried the first two women to raft through the canyon.

In 1949 600,000 visitors viewed the canyon, 12 went down the river, and this same year saw the 100th person to complete the river trip through the Grand Canyon. By 1954 the second 100 persons made the trip and in 1960 the first power boat travelled up the Colorado River through the Grand Canyon.

In 1963 the Glen Canyon Dam closed off the Colorado River and controlled the water discharge downstream. This changed the nature and personality of the river. By 1964 rafting was big business. Over 160,000 people made the river trip through the canyon.

In 1965 plans were proposed to construct two new dams within the Grand Canyon but public opposition prevailed. In 1966 electric power arrived at Phantom Ranch replacing the noisy generators. A trans-canyon water pipeline from Roaring Springs near the head of Bright Angel Creek was completed in 1970 and provided drinking water for all facilities on the North and South Rims and at Phantom Ranch and Bright Angel Campground at the canyon bottom.

In the 1970s there were several types of boat trips down the Colorado River through the Grand Canyon. There were the small specially built rowboats (dories) that held an oarsman and four to six passengers which rafted the canyon in about three weeks. There was also the newer pontoon boats powered by outboard motors that could carry up to 30 passengers and make the canyon trip in eight to 10 days. Trips began at Lee's Ferry and ended at Lake Mead.

The Grand Canyon was mapped in detail by the U.S. Geological Survey between 1970 and 1977, using modern surveying equipment and helicopter access.

In 1977 about 3,000,000 tourists visited the canyon: 250,000 hiked into the canyon, 19,000 rode mules to the bottom of the canyon, and 250,000 viewed the canyon from the air.

The Grand Canyon Railway from Williams to the South Rim was reinstated in 1989, with 99,942 passengers in 1992.

In 1991 an agreement was reached between the Grand Canyon Trust and the owners of the Navajo Generating Station which reduced sulfur emissions by 90%.

Today the South Rim is open all year, but because of heavy snows the North Rim is only open from about mid-May until late October.

River trips are limited to 22,000 persons each year and each trip must carry their own waste out. More than four million tourists from all over the world visit the Grand Canyon each year to hike, river raft, tour the canyon from helicopters and planes, and view and photograph the magnificent beauty of it all.

The Grand Canyon National Park Service is addressing problems that include air pollution, noise from helicopters and fixed-winged aircraft, controlled Colorado River flow and ever-increasing tourist numbers.

THE VISIONARIES

Eight men and one woman with vision helped shape the early understanding and development of the Grand Canyon and the Colorado River.

John Wesley Powell was the first man to conquer the Colorado River.

John Hance, William Wallace Bass, Louis Boucher, Ralph Cameron and Pete Berry all arrived at the south rim of the Grand Canyon in the 1880s. They came as prospectors, but soon realized that there was more money to be made in tourism. Each claimed their own territory along the south rim and developed tourist facilities and trails into the canyon.

Emery and Ellsworth Kolb opened their photography business on the South Rim in 1903, operating continuously for over 70 years.

Mary Jane Elizabeth Colter, an interior designer and architect for the Fred Harvey Company between 1902 and World War II, decorated and designed many of the buildings on the South Rim.

These courageous individuals paved the way.

John Wesley Powell

John Wesley Powell was born in 1834 in Mount Morris, New York. His father, a Methodist preacher, was vigorously against slavery. The family moved to Jackson, Ohio when John was a small child. Young John was frequently stoned by his classmates probably because of his father's stand against slavery and finally had to be schooled by a neighbor, a farmer and self-taught scientist. When John was 12 the family moved to a farm in Wisconsin where John assumed management of the farm while his father traveled. When John was 18 he began teaching in a one-room country school to earn money for college and for the next seven years he taught, attended colleges in Illinois where he received Bachelor and Master of Arts degrees, and explored the Midwest.

In 1856, at age 22, he went alone down the Mississippi River from the Falls of St. Anthony to its mouth. In 1857 he rowed the entire length of the Ohio River.

In 1858 he began teaching at Hennepin, Illinois and in 1860, at the age of 26, became superintendent of its schools.

At the outbreak of the Civil War in May, 1961, Powell enlisted in the Union Army as a private in the 20th Illinois Infantry and was soon promoted to sergeant major. When the group was mustered into federal service a month later he was commissioned a second lieutenant.

In November 1861 General Grant allowed Powell a short leave so he could travel to Detroit to marry his cousin, Emma Dean. By the end of the year Powell was made captain of Battery F, 2nd Illinois Artillery Volunteers.

On April 6, 1862 at the Battle of Shiloh, Powell was struck by a Minie ball in his right arm and the arm had to be amputated below the elbow. He returned to service when his wound healed and took part in the battles of Champion Hill and Black River Bridge. General Grant gave Powell's wife permission to accompany her husband on the battlefield to care for him. In 1865 Powell was mustered out of the service with a rank of lieutenant colonel, although he preferred to be called major.

Powell accepted the position of professor of geology and curator of the museum of the Illinois Wesleyan University at Bloomington. He often took his students into the field to collect fossils, minerals, plants and to observe animals. In 1867 he took a party of students to the Rocky Mountains to collect museum specimens. During the summer of 1868 Powell, his wife, 20 neighbors and students returned to Colorado and the Rocky Mountains. They built cabins and stayed through the winter, and it was during this time that Powell began to formulate his idea of exploring the Colorado River and the area of the Grand Canyon.

Powell had heard stories of earlier expeditions down the Colorado River that had failed and the explorers perished. He read of Lieutenant Joseph Ives who explored the Colorado River below the Grand Canyon and believed "that the Colorado, along the greater part of its lonely and majestic way, shall be forever unvisited and undisturbed."

Powell studied the few reports on the Colorado River and Grand Canyon, talked with Indians, hunters and mountain men who had crossed the area. He made the decision that it was possible to explore the Colorado by descending the river in small boats.

Powell received funding from private sources and the Illinois State Natural History Society. He was permitted to requisition military stores and had four boats built in Chicago and shipped to the expedition starting point at Green River Station, Wyoming Territory.

Powell's crew included his brother, Walter, and eight mountain men experienced in living off the land. There were no photographers or illustrators.

The expedition was launched on May 24, 1869 on the Green River. On June 9, well before reaching the Grand Canyon, one boat was lost to the rapids with most of the 10-month supply of provisions. Luckily no lives were lost. Shortly thereafter, one of the crew became discouraged and left, walking to a nearby settlement.

On July 21 they reached a portion of the Colorado River which flowed rough. They ran some of the rapids and portaged others. On August 3 they reached a beautiful canyon which they named Glen Canyon. On August 9 they reached another canyon whose walls were made up of many colors of marble. They named this canyon Marble Canyon.

On August 10 they reached the mouth of the Little Colorado River and camped there for three days. While there, Powell wrote the following in his journal:

> We have an unknown distance yet to run; an unknown river yet to explore. What falls there are, we know not; what rock beset the channel, we know not; what walls rise over the river, we know not.

On August 13 Powell and his expedition set off into the Grand Canyon, and on August 14 Powell describes the river in his journal:

> The canyon is narrower than we have ever before seen it; the water is swifter; the walls are set with pinnacles and crags; and sharp angular buttresses, which, bristling with wind- and wave-polished spires, extend far out into the river. As we proceed, the granite rises higher, until nearly a thousand feet of the lower part of the walls are composed of this rock.

About 11 o'clock we hear a great roar ahead, and approach it very cautiously. The sound grows louder and louder as we run, and at last we find ourselves above a long and broken fall, with ledges and pinnacles of rock obstructing the river. There is a descent of perhaps 70 or 80 feet in a third of a mile, and the rushing waters break into great waves on the rocks, and lash themselves into a mad, white foam. A portage would be impractical, and we must run the rapids or abandon the river. There is no hesitation. We step into our boats, push off, and away we go. ...

Running the Rapids on Powell's Second Expedition.   From Powell, 1875.

On August 15 Powell writes:

The day is employed in making portages and we advance but two miles.

On August 28 three of Powell's men would go no further because of the rough rapids encountered. They left the expedition on foot. Powell left one of his boats in case the three men decided to turn back, but it was later learned that they were killed by Indians.

On August 29 Powell and his remaining five crewmen emerged from the Grand Canyon near the stream called the Grand Wash. A day later Powell and his men reached the mouth of the Virgin River and were met by settlers.

Presumed dead after 99 days and more than 1,000 miles on the Colorado River, Powell and his men completed their epic journey.

Powell's expedition was only partially successful. Because of the loss of food and supplies it was hurried. Notes on the topography and geology were incomplete and only a few specimens were collected.

Powell planned for a second expedition. He returned a hero to Illinois and went on the lecture circuit to raise funds for the second expedition.

This time, planning an expedition that would last about a year and a half, he obtained funding from Congress, cached supplies along the river and engaged a surveyor, experienced photographer and illustrator. Preparations were completed by the spring of 1871 and on May 22 the second expedition set off from Green River Station. It took 4.5 months to reach the mouth of the Paria River at Lee's Ferry at the mouth of Glen Canyon.

Powell and his men camped at Lee's Ferry from September 1871 until August of 1872. During this time Powell traveled on horseback between the river and Salt Lake City exploring the canyon lands and studying the Indian tribes. The surveyor was mapping the area.

In August 1872 Powell and his men left Lee's Ferry and traveled down the river for the Grand Canyon. The river was high, swift and dangerous because of torrential rains and a heavy snowmelt in the Rocky Mountains. Because it was nearly impossible to control the boats in the rushing current, Powell halted the expedition when they reached Kanab Canyon.

Data obtained from the second expedition resulted in a topographic map of the Grand Canyon region. Powell's journal was used to publish his "Exploration of the Colorado of the West and It's Tributaries" in 1875. Hundreds of photographs were taken, many in stereoscopic views.

In 1878 Powell published, as a congressional document, his "Report on the Lands of the Arid Region of the United States," which included the physical characteristics of the land and rainfall, the need for a land classification system and drafts of proposed legislation providing for the organization of irrigation and pasturage districts.

In 1880 Powell was appointed director of the Smithsonian's Bureau of Ethnology, a federal agency which collected information on the disappearing Indian tribes of North America. Powell remained as director for the remainder of his life.

Because of Powell's background as a geologist, in March 1881 he became director of the U.S. Geological Survey when its first director, Clarence King, resigned. He expanded its geologic studies and topographic mapping throughout the country and promoted studies of soil, ground water, rivers, flood control and irrigation.

On June 30, 1894 Powell resigned as director of the U.S. Geological Survey because of poor health. His health declined and he died from a cerebral hemorrhage at his summer home in Haven, Maine on September 23, 1902 at the age of 69, survived by his wife, Emma, and only child, Mary Dean.

"Captain" John Hance

John Hance was described as one of the colorful characters found in Grand Canyon history.

Hance, possibly the first white settler in the canyon, arrived in 1883 and began exploring and prospecting the eastern part of the Grand Canyon.

An old animal or Indian trail into the canyon was improved which was later called the "Old Hance Trail." In 1884 Hance acquired squatter's rights to a ranch near his trail and in 1886 opened his ranch to tourists, constructing a cabin and tents. He operated a small business taking people down his trail to the Hance Rapids along the Colorado River. During this time he became a master story teller, similar to Will Rogers and Mark Twain.

During this time he befriended several other prospectors and, discovering asbestos, they located several claims in Grapevine Canyon south of the Colorado River, plus additional claims in the Red Canyon area in 1891.

Between 1892 and 1894 they located mining claims covering most of the tributary canyons on the north side of the Colorado River.

In 1894 Hance and his partners located and claimed a toll road 27 miles long which began on the south rim to the bottom of the canyon. This road crossed the Colorado River near the bottom of Hance Ranch (mile 77) and continued westward on the north side to Clear Creek about 3/4 of a mile east of the Hance (Red Canyon) Trail.

Camps were built near the asbestos outcrops and a small mining operation began. Ore had to be ferried across the Colorado River in a small canvas and wooden boat, which was always a challenge.

In 1895 Hance sold his ranch and trail to the Atlantic & Pacific Railroad Company which was renamed the Grand Canyon Hotel. Hance returned to mining his asbestos claims. Sixteen of his claims were patented in 1901 and were sold a few months later for $6,500. He remained as superintendent until 1904.

Hance then managed a small hotel and guided parties into the canyon for a short time until the business failed because it could not compete with the Fred Harvey Company and the Sante Fe Railroad's facilities at Grand Canyon Village. The Fred Harvey Company hired Hance to tell his stories to the tourists, which he enjoyed doing until 1918 when he became ill.

Hance died on January 6, 1919 and was buried in the cemetery at Grand Canyon Village.

William Wallace Bass

William Wallace Bass was born in Shelbyville, Indiana on October 2, 1849. His father left for the California Gold Rush where he died of Yellow Fever. The family then moved to New Jersey where Bill attended school through the Sixth Grade. He learned carpentry and studied telegraphy and at age 17 got a job on the Erie Railroad. During the next ten years his health deteriorated and his doctors suggested he should move to the arid southwest.

At age 27 Bill headed west and spent the early 1880s in New Mexico, Arizona and even Mexico. He wandered north in July 1883 to Williams, Arizona where for two months he worked at odd jobs doing house construction, playing the fiddle for dances and as a deputy sheriff. His health improved and with savings he attempted a little cattle ranching north of Williams.

Havasupai Indians guided Bass to the rim of the Grand Canyon in late 1883 and by early 1884 he erected a small homestead cabin near Havasupai Point. From there he could explore and prospect the canyon. Bass had heard a story about John Doyle Lee's (Lee's Ferry) cache of three five-gallon cans filled with gold hidden somewhere in the canyon.

Supai Indians showed Bass a spring below the rim and he improved an old Indian track into a trail down the upper canyon wall to what he called Mystic Springs, all the while searching for Lee's Cache.

By the spring of 1885, Bass figured that the Grand Canyon had great potential as a tourist attraction. At his homestead cabin he established Bass Camp which, in time, served photographers, artists, writers and geologists. Bass built another camp at Mystic Springs which became a base for visitors exploring the lower parts of the canyon.

He attempted, but was unsuccessful, to get the Atlantic & Pacific Railroad to extend the line to Bass Camp.

About this time he began using the title of "Captain" with his name.

For the next five years he explored the canyon and in 1890 located several asbestos claims on the north side of the Colorado River. He established a field camp, garden and orchard along the north side of Shinumo Creek. He built a rock cabin and packed in lumber on burros to build a crude wooden boat to cross the Colorado River at a place he called Bass Ferry. He also worked on a trail leading to the north rim.

During this time Bass discovered a fair showing of copper sulphide ore with some lead and silver in fissure veins in the Vishnu Schist near the canyon bottom. Over the years Bass hauled enough ore on burros from his mine in Copper Canyon to fill two railroad cars.

Eventually his camp and river crossing would be used by cross-country hikers along the only route from rim to rim available at that time.

During the late 1880s the route from Williams to Bass Camp on the canyon rim was a little-used wagon trail meandering through high grass. In 1891 Bass constructed the first rough road to his camp and purchased an old four-horse coach. He ran two weekly trips from the railhead at Williams.

In 1892 he built 35 miles of stage road from Ash Fork and opened a second stage line.

In late 1892 Bass met a 25-year-old music teacher, Ada Diefendorf, from New York who was touring the canyon and in January 1895 they were married. She became the first white woman to raise a family on the canyon rim. In addition to being a frontier wife, she became a hosteler, guide, hostess, cook, laundress, seamstress and chambermaid, all this time raising three daughters and a son. Periodically she loaded laundry onto a burro and made a rugged 11-mile trip down to the river and camped there overnight while doing the wash.

In 1898 catering to increased tourism, Bass extended an old Indian trail seven miles from Mystic Springs to the Colorado River.

When the Santa Fe Railroad built its branch line from Williams to the Grand Canyon Village in 1901, a flag-stop (Bass Station) was established 4.5 miles south of the village. Bass discontinued both stage lines and passengers traveling to Bass Camp would transfer from the train to horseback, wagons or stagecoach for the remainder of the trip. Departing tourists would have to flag down the train as it passed on its return to Williams.

In the late 1890s Bass and Ada moved to New York where their first two children were born. For the few years in New York, Bass lectured with lantern slides about the Grand Canyon. Returning to the Grand Canyon in 1906, Bass constructed a cable crossing over the Colorado River near his mine. The cable was improved in 1908 and that same year was replaced by a cage ferry. The wooden cage was large enough to carry pack animals one at a time, and was hung from four main cables attached to a pull cable. He ferried tourists, animals, hunting parties and asbestos from his mines.

In 1908 Bass, his son Billy, and his hired help packed out about 25 tons of ore on burros. Ninety pounds of high-grade handpicked asbestos was carried by each burro to the canyon rim, some of which was shipped to France to be used in the world's first fireproof theater curtains.

Overall, Bass's mines were not profitable and he made his money in the tourist business. He guided tourists west to Cataract Canyon and east to Grandview and Desert View. His daughter, Edith, helped guide tourists on the trails.

In 1911 he helped establish the first school at the canyon. He also erected a 12-room Bass Hotel at Grand Canyon Village and was ordered to remove it when the South Rim concession was awarded to the Fred Harvey Company.

Bass was a friend of the Havasupai, lobbying on their behalf in Washington, D.C. He obtained a school house, teacher, post office and medicine for them.

In 1912 Bass bought a seven-passenger Studebaker automobile for $1,500 to be used to transport tourists. His biggest business year was 1915 when he grossed $21,000.

"Captain" and Mrs. Bass entertained their last paying guests on September 15, 1923. During the 36 years Bass Camp operated, famous visitors included writer Zane Grey, artist Thomas Moran, naturalist John Muir, and industrialist Henry Ford. They soon after moved to Wickenburg, Arizona.

Early in 1926 the Santa Fe Land Development Company purchased all of Bass's interests in the Canyon for $25,000.

In Wickenberg Bass gardened, wrote, prospected and operated a small camp for tourists.

During his 35 years at the Grand Canyon, Bass constructed more than 50 miles of trails in the canyon, including the first cross-canyon trail. Captain Bass died in 1933 at the age of 84. Mrs. Bass died in 1951.

In the early 1960s his cables across the Colorado River were removed because they were a danger to low-flying aircraft.

Today all of the asbestos workings in the Grand Canyon are closed to visitation by order of the National Park Service.

Louis Boucher

Louis Boucher was French-Canadian, born in Sherbrooke, Quebec. He roamed the Grand Canyon and was known as "The Hermit." Hermit Basin, Hermit Canyon, Hermit Rest, Hermit Shale and the Hermit Fault are all named after him.

He settled at Dripping Springs, about 1,000 feet below the south rim and about eight miles west of present-day Grand Canyon Village.

He built the Boucher Trail around Columbus Point and into Boucher Canyon where he planted an orchard and garden.

In 1889 Boucher opened a tourist facility at Dripping Springs. This was the first camp for tourists below the south rim on the Tonto Platform near Hermit Creek. Boucher also prospected, discovering a small copper deposit in the Vishnu Schist near his cabin.

Louis Boucher left the Grand Canyon after 1909, and little remains of his presence in the canyon.

Ralph Cameron

Ralph Henry Cameron was born in 1864. In 1883 he left Boston at age 19, arriving in Flagstaff with his brother Niles on the new Atlantic & Pacific Railroad. For a few years Cameron operated a nearby sheep ranch and mercantile business.

In 1890 he and other prospectors filed numerous placer mining claims within the Grand Canyon. They also improved an old Havasupai trail covered by the claims, calling it the Cameron Trail which would later become the Bright Angel Trail.

In 1891 Pete Berry, Cameron's partner, obtained a franchise to operate the trail as a toll road. The toll was one dollar per footman and horse or mule and rider.

It should be noted that in those early years the law allowed anyone to build a trail or road on public domain or through their own mining claim without government supervision.

Between 1897 and 1903 Cameron built a small hotel on his south rim placer claim.

The Santa Fe Railroad spur line between Williams and Grand Canyon Village was completed in 1901, and the Santa Fe completed the El Tovar Hotel on the South Rim in 1905. There would be competition and rivalry with Cameron for the tourist business.

Bright Angel Trail was the closest and most accessible route into the canyon from Grand Canyon Village and the train station. Because the tourists and guides were reluctant to pay steep toll prices, the Santa Fe Railroad and Fred Harvey Company began building their own trail at Hermits Rest, eight miles west of the village.

Cameron knew that the owner of a placer mining claim had surface rights and anyone going on the claim would need the owner's permission, or could be deemed a trespasser. Not to be outmaneuvered, in 1905 Cameron located 39 placer claims covering 13,000 acres over the Santa Fe station grounds, the site of the El Tovar, and various rim points. The claims also covered Indian Garden, the only water source along the trail which he controlled as a toll road, and the section of the trail along the Colorado River.

In 1906 Berry's franchise to operate the trail expired and control of the trail reverted to Coconino County. In early 1907 Cameron located an additional 28 claims covering the trail below Devil's Corkscrew and also acquired the trail franchise from friends.

The Santa Fe Railroad began a legal challenge on the validity of Cameron's mining claims, alleging that there were no economic minerals on most of the claims and that the claims were not valid. Litigation would ensue for years.

In 1912 a Philadelphia syndicate took over seven of Cameron's placer claims near Indian Garden. The 28 claims located below the Devil's Corkscrew portion of the trail were bonded to New York investors who planned to placer gold.

In 1916 the Santa Fe Land Company acquired Cameron's mining claims covering the upper portion of the Bright Angel Trail for $40,000.

The Grand Canyon National Park was established in 1919, and the National Park Service rebuilt the Bright Angel Trail to bypass the steep Devil's Corkscrew on Cameron's claims. The same year Cameron became a United States Senator from Arizona.

Cameron and his partners still controlled Indian Garden and the mining claim at the top of the Bright Angel Trail, but he was never able to patent his placer mining claims because economic mineral deposits did not exist on his claims. Finally the claims were declared invalid, but Cameron refused to vacate them.

In 1923 Cameron was the defendant in several suits filed in the U.S. District Court in Arizona by the U.S. Government and several businessmen to invalidate his mining claims and for contempt because he failed to remove structures and employees from Indian Garden and the mining claim at the head of Bright Angel Trail.

Cameron lost his reelection bid to the U.S. Senate in 1926 and gave up the fight to control his Bright Angel Trail. In 1928 the title to the Bright Angel Trail passed to the National Park Service.

Cameron moved to Philadelphia and died on February 12, 1953. He is buried in the cemetery at the Grand Canyon.

Pete Berry

Pete Berry arrived at the Grand Canyon in 1888. Exploring the canyon on foot and horseback, he met other prospectors and acquired a desire to strike it rich. In 1892 and 1893 Berry and his partners built the Grand View Trail down to his mine. A few years later the Grand View Hotel was built and tourists were guided down the Grand View Trail. Berry's partners included Ralph and Niles Cameron and in 1895 they formed the Grand Canyon Copper Company.

In 1902 the Grand Canyon Copper Company's assets, including the mines, mill site, trail and the Grand View Hotel, were sold to a consortium of capitalists from New England and Flagstaff.

In the early 1900s copper ore was brought up the Grand View Trail on strings of eight to ten mules, each carrying about 200 pounds and making a trip and a half per day. Ore was stockpiled on the rim and then transported to the Santa Fe spur line that ran to the Grand Canyon Village 14 miles to the west, and from there 600 miles by rail to the smelter at El Paso, Texas.

There were plans to build a dam for electric power generation on one of the tributaries of the Colorado River and also an aerial tramway to haul ore to the rim, but in 1907 the price of copper plunged from 24.5 cents to 13 cents per pound and mining operations ceased.

The Grand View Mine produced about $75,000 in copper. The properties were sold to William Randolph Hearst in 1913, and Berry and his wife remained as caretakers until 1919 when they moved to the ranch of their son Ralph.

Martha died in 1931 and Pete died in 1932. Both are buried in the Grand Canyon cemetery. The National Park Service bought the Hearst properties in 1939 for $85,000 and removed the rim buildings in 1959.

The Kolb Brothers

Ellsworth and Emery Kolb played a big part in the development of the Grand Canyon into a major tourist attraction in the early 1900s.

The brothers were born in Smithfield, Pennsylvania (1876 and 1881 respectively) to a Methodist minister and his wife.

Ellsworth headed west in 1900 at age 24, and worked a number of laborer jobs in Colorado, Wyoming and California. He obtained employment as a carpenter with the Santa Fe railroad in late 1901, and first set eyes on, and fell in love, with the Grand Canyon. During these two years Emery, at home in Pennsylvania, bought a small camera and experimented with photography.

Ellsworth briefly returned home and, with his tales of the wonders of the Grand Canyon and the business potential of photography at the Grand Canyon, he encouraged his brother to join him. Upon his return to the Grand Canyon in October 1902 Ellsworth wrote Emery that he had found him a job in an asbestos mine.

While waiting to transfer trains at Williams, Emery visited a photography store and conversed with its owner, who told Emery his business was for sale. Emery continued to the South Rim where he learned that the asbestos mine had shut down so there was no job waiting for him.

Ellsworth and Emery agreed to buy the photography business in Williams for $425 with a year to pay it off. Within the next year they became proficient in photography. Their main source of income was pictures of saloon girls, but on weekends they prowled the canyon with a camera.

The Santa Fe Railroad Company arrived at the South Rim of the Grand Canyon in late 1901 and brought the Fred Harvey Company with them. Their goal was to completely control the development of all tourist business.

The Kolb Brother's goal was a photography studio at the Grand Canyon. The railroad company, with its money and power, objected to the studio; consequently, the Forest Reserve Officer refused permission to open the studio. Hostilities with the railroad and Fred Harvey Company lasted for years.

In 1903 the Kolbs talked to Ralph Cameron about setting up their studio on his mining claim at the head of the Bright Angel Trail. Cameron agreed and the brothers constructed a tent building and store on Cameron's mining claim.

Photographs were taken of tourists on mules or hiking down the trail into the canyon, and Emery would run 4.5 miles down Bright Angel Trail to Indian Garden, the nearest source of water. Covering the entrance to one of Cameron's adits with a blanket, Emery developed the glass photos and prints, and then would run back up the trail to the rim to sell the pictures to the returning hikers and mule riders. He often made the trip twice daily.

The little tent studio became inadequate by 1904, and Cameron granted the Kolbs permission to build a small two-story frame building that extended down the side of the canyon at the head of the Bright Angel Trail. The new studio was completed in 1906.

The brothers soon had enough money to build a small studio for developing their film at Indian Garden. Later, mules would carry water up from Indian Garden to develop the film at their studio on the rim.

On October 17, 1905 Emery married Blanch Bender of Prescott.

The Kolb Brothers made their first trek into the Little Colorado River Gorge in September 1906. Packing their bulky cameras, they took spectacular photographs. Their second, a longer trip into the Gorge, was in 1909.

In 1907 the Kolb Brothers explored Havasu and Cataract canyons, home of the Havasupai Indians, and they took the first photographs of that area.

In June 1907 Emery's daughter, Edith, was born.

During the winter of 1910 the Kolb Brothers made plans to retrace the river trips of John Wesley Powell of 1869 and 1871. In 1911 they ran the Colorado River from Green River, Wyoming to Needles, California and covered about 1,200 miles, 365 big rapids and many smaller ones, in 101 days. They took the first motion pictures of the canyon, which played to packed houses everywhere it was shown. The brothers then expanded their studio and added a small theater to show their movie.

In 1912 the brothers went on lecture tours all over the United States, showing their moving pictures of the Grand Canyon.

In 1913, with the permission of the Department of Interior, they took moving pictures of Hopi rituals.

That same year Ellsworth took a river trip from Needles to Mexico.

In 1914 Emery was drafted into the Army at the start of World War I as a First Lieutenant in the Photographic Section of the Signal Corps.

In 1915 the brothers disagreed on the operation and direction of their business and subsequently they agreed to work the canyon separately in two-year shifts.

The August 1919 issue of *National Geographic* featured the Kolb Brothers. Also in 1919 Emery was sent to Alaska as part of a National Geographic team to photograph an active volcanic area, the Valley of Ten Thousand Smokes. He returned and took over his two-year shift at the canyon.

In 1921 the Kolb Brothers were employed by the U.S. Geological Survey as guides and boatmen to explore the upper canyons of the Colorado River from Green River to Lee's Ferry.

On August 8, 1922 Ellsworth, with an interest in aviation, flew as a passenger in a biplane from Williams, Arizona over the rim of the Grand Canyon taking moving pictures. The plane landed on a plateau below the rim near Bright Angel Trail with a little damage, and was able to take off successfully the next day.

By the end of 1922 the relationship between Ellsworth and Emery continued to deteriorate, and they drew up an agreement dividing their business interests, including mining claims they had located over the years and lecture areas.

In 1923 Emery accompanied a Colorado River trip with the U.S. Geological Survey and California Edison personnel from Lee's Ferry that would survey the Grand Canyon in detail for the first time. The purpose of the trip was to select a dam site for flood control, hydroelectric power generation and water storage. This trip recommended Black Canyon as the site for the future Boulder Dam. On October 19th the boats arrived at Needles, California.

The brothers finally separated in 1924. Emery kept the Grand Canyon studio and house, and Ellsworth moved to Los Angeles and received $150 per month from Emery.

In 1926 the last addition to the Kolb Brothers' studio was completed. Emery continued to operate the studio and Ellsworth reappeared only briefly from time to time until his death in 1960. Emery would operate the photography business at the Canyon until his death in 1976 at the age of 95.

Covering a span of 74 years the Kolb Brothers' collection consists of 350 cameras and other equipment, 250,000 photographs and 40,000 feet of motion picture film. This does not include photographs sold to the tourists.

The Kolb Brothers' photographs of the Grand Canyon area scenery and tourists on mules down the Bright Angel Trail are a legacy to their pioneering spirit in developing the Grand Canyon into the tourist attraction that it is today.

The Desert View Watchtower, located 25 miles east of Grand Canyon Village. It was designed by Mary Elizabeth Coulter in 1932.   Photo by Ann Ettinger.

Mary Jane Elizabeth Colter

Born in 1869 in Pittsburgh, Pennsylvania, Mary Jane Elizabeth Colter was educated in interior decorating and architecture at the California School of Design in San Francisco, specializing in Mexican and Indian cultures of the Southwest.

The Fred Harvey Company had operated hotels, restaurants, shops and dining cars for the Santa Fe Railroad since 1876. In 1902 Colter contracted with the Harvey Company, and was commissioned to decorate the interior of the buildings along the Santa Fe route. Her first assignment at the South Rim of the Grand Canyon was to design and decorate the new Hopi House in 1905. She supervised the building of the Pueblo structure which was constructed by Hopi workmen. In 1910 she became a full-time employee of the Harvey Company as Company Architect and Decorator.

Her other accomplishments included Lookout Studio and Hermit's Rest in 1914, the Phantom Ranch in 1922, the Desert View Watchtower, 25 miles east of Grand Canyon Village, in 1933, and Bright Angel Lodge in 1935.

Other notable projects during her long career included the Alvarado Hotel and Indian Museum in Albuquerque, New Mexico (1902), the El Ortiz Hotel in Lamy, New Mexico (1910), The El Navajo Hotel in Gallup, New Mexico (1923), the La Fonda Hotel in Santa Fe, New Mexico (1927), the La Posada Hotel in Winslow, Arizona (1930), and the Painted Desert Inn at the Petrified Forest (1947).

Colter worked for the Harvey Company for over 40 years, never married, and died in 1958 at the age of 89 in Santa Fe, New Mexico. Her architectural and interior design style incorporated the cultures of the Spanish-Mexicans, Hopi, Zuni and Navajo Indians in the backdrop of the spectacular Grand Canyon.

One of a herd of Bighorn Mountain Sheep viewed by mule riders near the top of the Bright Angel Trail, November 2004. Photo by Ann Ettinger.

TRAILS, HIKERS, MULES and RIDERS

There are more than 19 trails into the Grand Canyon, originally made by deer, bighorn sheep, Indians and prospectors. Today most are not maintained and have been abandoned.

The National Park Service maintains only the Bright Angel and Kaibab trails. The Bright Angel Trail begins near Grand Canyon Village and drops some 4,500 feet to the Colorado River in about eight miles, where it joins the Kaibab Trail. The trail averages five feet in width and mules have the right-of-way.

The South Kaibab Trail begins near Yaki Point about four and a half miles east of Grand Canyon Village and descends about 4,800 feet in seven miles where it joins the Bright Angel Trail and crosses a suspension bridge (The Black Bridge) over the Colorado River. The North Kaibab Trail follows Bright Angel Canyon about one-half mile to Phantom Ranch and then northward 13 miles up to the North Rim, ending near Grand Canyon Lodge at an elevation of 8,200 feet. Constructed between 1924 and 1928, part of the trail was built using jackhammers and many tons of dynamite.

Other trails include the Hermit Trail, Grandview Trail, Bass Trail, Boucher Trail and Tonto Trail.

Hikers who want to travel the abandoned trails must register at a ranger station, file a "hike" plan and submit to an equipment inspection. The ranger may refuse permission to hike closed areas.

A number of abandoned trails lead to some of the 83 old mining claims in the canyon. Old mines yielded small tonnages of asbestos, copper, silver, lead and bat guano. All the mines had nearby camps, all of which are closed areas.

Park rule is "if you pack it in, you pack it out," and it applies to everything.

Hikers and mule riders view the remains of ancient Indian occupation, innumerable plant species and a variety of animals including mountain sheep and burros. The burros are descendants of animals turned loose by miners and prospectors in the late 1800s.

Mules have been mentioned in Egyptian scrolls and the Bible. They were used by Alexander the Great and Napoleon.

Mules are the result of a cross between a male burro (jack) and a female horse (mare). Male mules are called Johns and females Mollys. Mules have the build of the horse and the disposition of the burro. A mule can do three times as much work as a horse on a third of the feed, and can and will eat anything. Riding a mule is much smoother than riding a horse and a mule is more sure-footed.

The first mule ride operation from the Grand Canyon Village area began in 1904 by the Fred Harvey Company and has operated continuously for more than 100 years.

Grand Canyon mules are considered the best in the world. Eight to 12 mules are bought at auction each year. They are judged on being smart, big-hearted, gentle, calm and not being easily spooked by noises. Maybe one in 1,000 passes the test.

After arriving at the Grand Canyon the mule spends several days in the corral getting to know new surroundings and acclimating to the 6,000 feet elevation. And then begins training by first packing groceries and other supplies down to Phantom Ranch and the nearby Ranger Station. On the return trip they pack garbage back to the rim.

After several months as pack mules and, now accustomed to the trails, the mules are ridden by the wranglers (guides) in and out of the canyon, first with the pack train and then leading tourist groups. After more months they may finally graduate to the string used for the mule riders.

The wranglers know the personality of each mule, and the trail boss tries to match each mule with the individual rider.

Mule riders must be at least 4'7" in height (children), in good health, not weigh over 200 pounds fully clothed and be fluent in English. And they do check!

About 160 mules are available for use. Most mules are from five to 25 years old, work five days a week and have a working life of 15 to 20 years.

When mules, who often reach the age of 25 to 30 years, become too old to safely cope with the trail, they are retired and eventually sold at auction to someone who wants a pet for their children or a nice mule for pleasure riding.

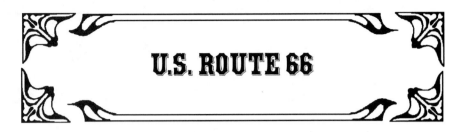

U.S. ROUTE 66

U.S. Route 66 was has been called many names: "The Main Street of America," "The Wine Road," "The Will Rogers Highway," and "The Mother Road."

Route 66, in general, had four periods of evolution. First was its alignment, assigned in 1926, following the National Old Trails Highway in northwest Arizona. Second were the improvements in the 1930s that shortened distances and widened the still unpaved road. Third was the paved road period between 1938 and the late 1950s, with more alignment changes. And finally was the era of construction of Interstate 40 in the late 1950s through 1984 when the final bypass occurred at Williams.

In recent years Route 66 has achieved a historic and romantic status that is appealing to many travelers. Its history has been documented in newspapers, books, magazines, songs, radio shows, television specials and road races. The more famous referrals include:

- John Steinbeck's "The Grapes of Wrath," published in 1939. This book vividly describes the migration westward of desperate families fleeing the depression in Oklahoma to a new life in California along Route 66.
- Bobby Troup's song, "Get Your Kicks Out of Route 66," sung by Nat King Cole in 1946, hit the top of the charts.
- A successful television series in the early 1960s used Route 66 as its theme for George Maharis and Martin Milner and their Corvette.

It wasn't until the late 1980s that local groups banded together and advertised, "slow down, see and enjoy Old Route 66."

Before U.S. Route 66

The earliest east-west travel routes across what is now northern Arizona were old Indian foot trails which were later used by the early explorers. Called the Rio Grande-Pacific Ocean Trail, the Hualapai section was a main travel route used for centuries by Indians as a trade route along which goods moved between the Pueblo peoples of what today is New Mexico, Indians of the Pacific Coast and all tribes between. The trail, going westward, crossed the Hualapai Valley through Railroad Pass, then to Sitgreaves Pass and down into the Mohave Valley.

In 1776 Fray Garcés, with Indian guides, traveled eastward along the Rio Grande-Pacific Ocean Trail, as did early explorers.

In 1852 Captain Lorenzo Sitgreaves and his party explored a possible wagon route along the 35th Parallel through northwest Arizona.

In 1853 Congress commissioned Captain Amiel Weeks Whipple of the Army Topographic Corps to conduct a survey for a proposed trans-

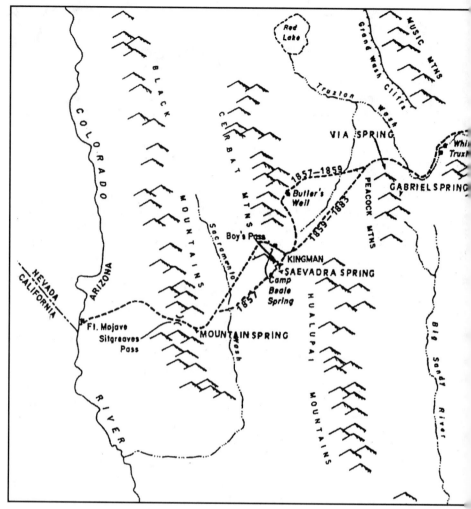

The Beale Wagon Road, 1857-1883.

continental railroad, but Congress eventually decided against the railroad and subsidized a network of wagon roads intended to improve military and civilian communication throughout the western frontier.

In 1854 Francois Xavier Aubrey was the first man to take a wagon from San Jose, California along the 35th Parallel across northwest Arizona and then to Santa Fe, New Mexico following Sitgreaves' survey.

In 1857 Edward Beale, a U.S. Navy Lieutenant attached to the U.S. Army, was commissioned by the U.S. Government to blaze and build an all-weather wagon road from Fort Defiance in the New Mexico Territory (now eastern Arizona) to the Mojave River along the 35th Parallel, essentially following the well-traveled Rio Grande-Pacific Ocean Trail.

A part of the project was the "camel experiment," a study funded by the U.S. War Department to test the use of camels to transport supplies and mail in the southwest desert. The camels performed successfully in the desert environment; however, the project was abandoned for a variety of

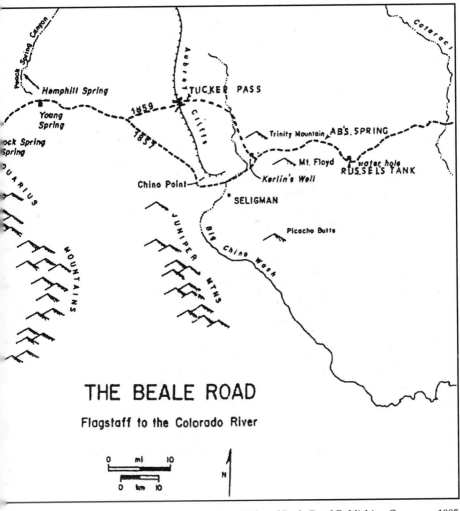

THE BEALE ROAD

Flagstaff to the Colorado River

From Tales of Beale Road Publishing Company, 1985.

reasons, including the beginning of the Civil War and objections from teamsters who used mules and horses. This first federally-funded road became a popular route for prospectors seeking gold, silver and copper.

Ferries were established to cross the Colorado River at Hardyville and Fort Mohave so travelers could continue their journey westward or eastward.

In 1864 the incorporation of the Prescott and Mojave Toll Road was approved by the Mohave Legislative Assembly with tolls as follows: "each wagon and two horses 1-1/2 cent per mile and each additional animal 3/4 cent per mile; one horse and vehicle 3/4 cent per mile; pack animals 1/2 cent per mile; horned cattle, horses and mules and others in droves 1/2 of 1 cent per mile; sheep, goat, or hog 1/8 cent per mile." The toll road established a station located at Beale's Springs.

In the early 1880s Lewis Kingman surveyed a railroad route between Albuquerque, New Mexico and Needles, California which closely followed Beale's road for much of its length.

125

There was really no need for an automobile highway across northern Arizona until the automobile became plentiful enough to challenge the railroads for tourism and basic transportation in the West. In the early 1900s many groups lobbied for the building of roads suitable for automobiles.

The National Old Trails Highway

In 1910 there were 180,000 registered automobiles in the United States. The National Old Trails Association was formed in 1912. One of its goals was to link the United States from coast to coast by retracing America's historic trails. In Arizona the trails included part of Beale's Wagon Road to California, the Santa Fe Trail and the Grand Canyon Route of Matthew J. Riordan. In 1914 these trails were joined and called the National Old Trails Highway.

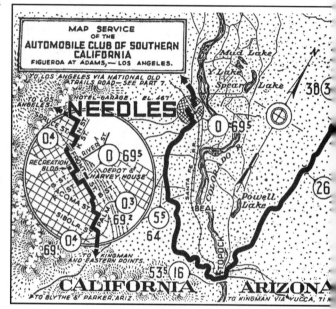

In northern Arizona most of the towns grew out of settlements that were along the railroad, and were originally built as stops where the early trains could obtain fuel, water and repairs. The earliest auto travelers drove on dirt tracks that became mud traps when it rained. These travelers crossed sections of sand, jagged rocks. They crossed streams and small rivers, and camped out when day was done or repairs were needed. They fueled up where they could and

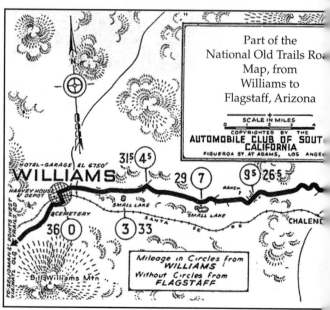

carried many gallon cans of extra fuel with them, along with oil, bags of water, spare tires and parts so repairs could be made easily when necessary. Fuel was usually found at a house with one or two fuel service pumps in front, in the railroad settlements or along the road.

At that time Kingman was the hub of a mining region and an important shipping point for cattle, sheep and wool. Flagstaff was an important center for the lumber industry and cattle ranching as well, and was also one of the gateways to the Grand Canyon.

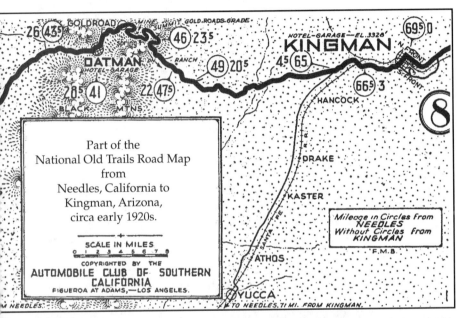

Part of the
National Old Trails Road Map
from
Needles, California to
Kingman, Arizona,
circa early 1920s.

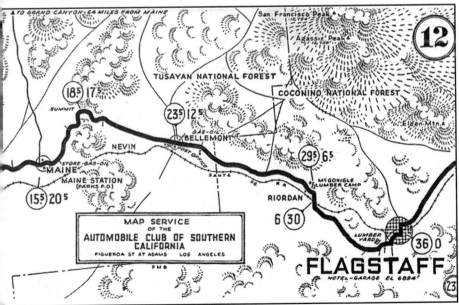

The alignment of the National Old Trails Highway followed the land, skirting around hills and other geographic obstacles. There was no heavy-duty earth-moving equipment, and the road meandered across the country and followed gentle grades whenever possible.

By 1916 a bridge was completed across the Colorado River at Topock, ending the need for a ferry crossing.

Because of the new popularity of the automobile and the miserable condition of most of the roads across the country, the first federal legislation for public highways was passed in 1916. The Federal Aid Road Act gave the states road construction funds along with a number of conditions:

- States could apply for assistance from the Department of Agriculture for funds to build and improve "rural post roads."
- The road route would currently, or likely in the future, carry the U.S. Mail between places of over 2,500 persons based upon the 1910 census.
- There was a road building cap of $10,000 per mile (later increased to $20,000, then down to $15,000 per mile), exclusive of bridge construction costs.
- Funds were distributed to states based upon their relative percentage of the total U.S. population.

The Federal Aid Road Act of 1916 was revised in 1921 and again in 1925, when legislation included a plan for national highway construction.

The construction and maintenance of sections of the National Old Trails Highway was an expensive challenge. The road shifted with every realignment of the Santa Fe Railroad, and west of Kingman a steep and winding section cut across the Black Mountains and through Oatman. Violent summer storms could wash away sections of the road.

The Federal Aid Highway Act of 1921 provided money for improvements on the National Old Trails Highway in Arizona when the federal highway system of numbering roads began.

U.S Route 66

By the 1920s there were more than 17 million cars, trucks and buses traveling on America's roads.

In 1925 Arizona reported a daily count of 338 cars along the National Old Trails Highway. The section between Ash Fork and Seligman was described as "a 20-mile per hour road, unimproved with boulders in the road that make it a menace to a low-clearance car."

In 1926 U.S. Highway 66 was the number assigned to the Chicago-to-Los Angeles road then known as the National Old Trails Highway in Arizona. Only about 800 of the 2,200 miles of Route 66 were paved, none in Arizona which were mostly dirt and gravel. It closely followed the Santa Fe Railroad across northern Arizona. The section over the Black Mountains near Oatman was one of the most feared grades on the entire route due to its twisting, rocky nature.

Automobiles and trucks of the 1920s, 1930s and 1940s were primitive by today's standards. Engines would overheat, tires would easily puncture

Route 66 near Sitgreaves Pass, Arizona, elevation 3,652', looking west.
Courtesy of the Mohave Museum of History and Arts, Kingman, Arizona.

and blow out, gears could break and brakes could fail. The winding, narrow and steep grades of Route 66 on both sides of Sitgreaves Summit, a few miles east of Oatman, were a true test for most cars and trucks. From the lookout just west of the summit you can still see at least one mangled car body down a steep slope.

In 1929 the 1,200-mile western portion of Route 66 from Texas to southern California had only 64 miles of surfaced highway, none in northwest Arizona with the exception of a few miles within the city limits of Flagstaff and Kingman.

By 1930 the trucking industry rivaled the railroad for a share of the American shipping business. Route 66 was a more direct way to travel between Chicago and the Pacific Coast, and enjoyed a better climate for winter travel than the northern highways.

The Great Depression of the 1930s brought a migration of some 210,000 people from the Dust Bowl of Oklahoma and Texas to California, many traveling Route 66. John Steinback, in his "The Grapes of Wrath," referred to U.S. Highway 66 as the "Mother Road," and others who traveled it thought of it as the "Road to Opportunity."

With cross-country automobile travel came the need for fuel, food, overnight accommodations and auto repair shops. The "auto camp" developed as townspeople along the road roped off spaces where travelers could camp at night. Camp supervisors provided water, fuel, wood, privies or flush toilets, showers and laundry facilities free of charge. It wasn't long before the "tourist home" — a private residence or outbuilding where travelers could share an overnight home environment and facilities for a few dollars and then be on their way the next morning — became popular.

The auto camp and tourist home evolved into the "cabin camp" or "cottages" that offered minimal comfort at affordable prices.

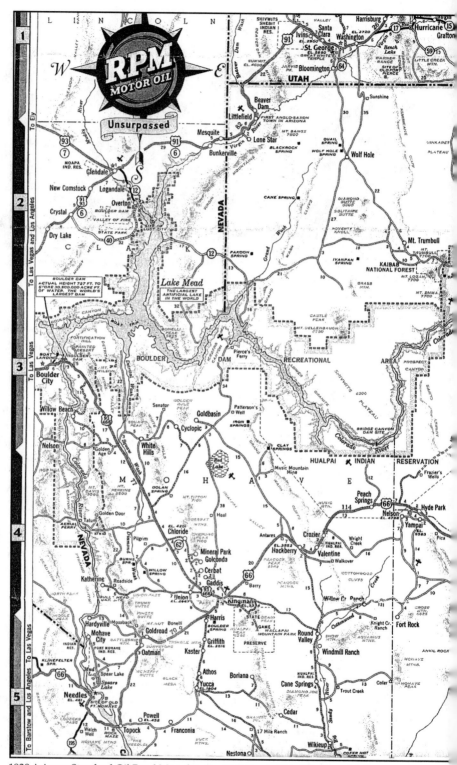

1938 Arizona Standard Oil Road Map showing U.S. Route 66 in Northwest Arizona.

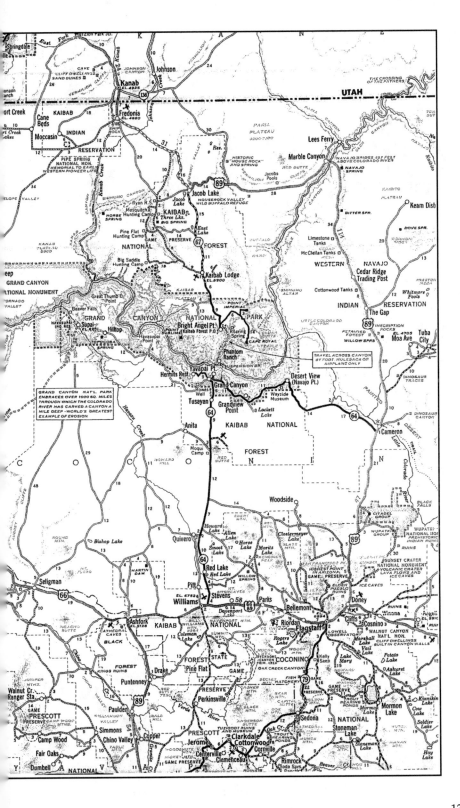

Roadside cabins at Cool Springs on Route 66 four miles east of Sitgreaves Pass, circa 1940. Courtesy of the Mohave Museum of History and Arts, Kingman, Arizona.

As traffic increased crude filling stations evolved, each associated with a particular petroleum company. Then came service bays, mechanics and tire outlets.

Between 1933 and 1938 thousands of young men were hired as laborers on road gangs to grade and pave the final stretches of Route 66.

In 1938 the entire length of the Chicago to Los Angeles Route 66 was finally declared "continuously paved with an all-weather capability."

And just in time for World War II.

The western United States, because of its isolation and dry climate, was ideal for military bases. With these numerous new military camps scattered along the road, many mile-long convoys of troops and military equipment traveled both east and west over Route 66, which consequently required much repair and repaving. During wartime repair costs were not an issue, and the road was kept in good condition even though highways of this period were inadequate for large volumes of traffic. Even with repairs, World War II truck traffic deteriorated the nation's highways. Most roads, including Route 66, were functionally obsolete and dangerous because of narrow pavements and old bridges with reduced carrying capacity.

Riverview Court on Route 66, seven miles west of Oatman, circa early 1940s. Courtesy of the Mohave Museum of History and Arts, Kingman, Arizona.

Heavy-duty earth-moving equipment was now available to realign difficult sections of Route 66, and modern engineering designs eventually yielded stronger highway structures.

Released in 1946 was a song written by Bobby Troup and sung by Nat King Cole with the words, "… get your kicks on Route 66." This song popularized the highway and there was a relative explosion of tourist travel that taxed its design and more realignments were made.

Route 66 provided access to many natural wonders in Arizona, including the Grand Canyon, Petrified Forest, Meteor Crater, and Anasazi and Pueblo ruins.

With the increase in auto travel the auto camps and cabin camps were replaced by motor courts where all the rooms were under one roof. Soon motor courts attracted adjoining restaurants, souvenir shops and swimming pools. The 1950s saw the motor court evolve into the motel, and strong competition yielded a variety of facilities and services at a wide-range of cost.

U.S. Route 66 through Oatman, Arizona, circa mid-1940s.
Courtesy of the Mohave Museum of History and Arts, Kingman, Arizona.

One afternoon in 1951 there was a steady stream of traffic entering Oatman on Route 66; the next day there was none. A new realigned section of Route 66 had opened between Kingman and Topock bypassing Oatman. Six of the seven service stations in Oatman closed and other businesses soon followed.

The Federal Aid Highway Act of 1956 approved a 42,500 mile modern national interstate highway system designed to connect 90 percent of all American cities with populations greater than 50,000.

Route 66 would be replaced by I-40 across northern Arizona in order to avoid delays, and straightening alignments to the total exclusion of many existing towns along the old highway.

In October 1984 the final section of the highway was replaced by Interstate 40 at Williams, Arizona. U.S. Highway 66 (Route 66) was decommissioned and returned to the status of local road.

In 1987 the Historic Route 66 Association of Arizona was formed by 15 people from Oatman to Seligman. The group expanded to membership across Arizona to promote the preservation of Historic Route 66.

Route 66 linked hundreds of towns and cities between Chicago and Los Angeles, forming the main street of many towns along its route. This prompted the nickname "Main Street of America."

Much of old Route 66 still exists, being used as local roads and for access by tourists who have time to explore. The remaining portions are a living history of the evolution of highway development in Arizona, from the primitive wagon trails to country roads, to the early federally subsidized highways, and finally to today's interstate highways. In some areas you can see two or three generations of road cuts.

Today in northwest Arizona, historic Route 66 can be traveled for 158 miles, beginning west of Ash Fork and continuing west through Seligman, Peach Springs, Kingman and Oatman to Topock along the Colorado River.

In the hills and canyons between Seligman and Hackberry you can still see remains of the graveled National Old Trails Highway which became Route 66 before realignment to the paved road. The old gravel roadway followed the natural contour of the land with few cuts in solid rock.

THE CIVILIAN CONSERVATION CORPS IN NORTHWEST ARIZONA
(1933-1942)

One of the most successful federal programs of the Great Depression era was the Civilian Conservation Corps (CCC). This program played an important role in improving parts of Northwest Arizona.

The CCC had three main purposes: to teach unskilled, unemployed men a trade; to provide paid employment; and to conserve natural resources. It was established by the United States Congress on March 31, 1933 as a part of the New Deal program of President Franklin Delano Roosevelt.

The CCC Act authorized an enrollment not to exceed 300,000 men at any one time, of which not more than 30,000 could be war veterans. The enrollees had to be unmarried male, U.S. citizens between the ages of 17 and 23 years, unemployed, and in need of a job. The enrollment was for a period of not less than six months and reemployment could not exceed two years. The pay of enrollees was $30 a month, assistant leaders $36, and leaders $45 per month. In addition to wages, the enrollees were provided with living quarters, subsistence and clothing, medical attention, hospitalization and transportation.

The program was designed as a quasi-military operation run by U.S. Army officers with a chain of command. Men, using tools rather than weapons, enlisted to enter the program and when their time was up, either reenlisted or were discharged.

Each camp had two Army officers commanding the camp. The camps had barracks, a mess hall with a kitchen, welfare building, infirmary, bath house and restrooms. There was also an incinerator, laundry boiler, garbage rack, pump and generator buildings, and a storage shed for gasoline. New wells and storage tanks provided water and each building was provided electricity from the generator. Each camp had its own sewage disposal system. Pickup and stakebed trucks provided transportation.

Part of the CCC program was to provide an opportunity for education and entertainment. Each camp provided classes in reading, writing, arithmetic and spelling. On-the-Job-Training included tractor operation, automobile mechanics, grader operation, carpentry, surveying, road construction, manufacturing adobe bricks, fence construction, small dam construction, and other trade work.

Transportation was provided to night classes if enough men were interested. Otherwise, the instructors would come to the camp.

Certificate of Discharge

from

Civilian Conservation Corps

TO ALL WHOM IT MAY CONCERN:

THIS IS TO CERTIFY THAT * __WILLIAM H. MC B████__, A MEMBER OF THE

CIVILIAN CONSERVATION CORPS, WHO WAS ENROLLED __April 22, 1936__ AT
(Date)

__Waxahachie, Texas__, IS HEREBY/ HONORABLY DISCHARGED THEREFROM, BY REASON

OF ** __To accept employment that will better his condition__

SAID __WILLIAM H. MC B████__ WAS BORN IN __Mountain Peak__

IN THE STATE OF __Texas__ WHEN ENROLLED HE WAS __26__ YEARS

OF AGE AND BY OCCUPATION A __Farmer__ HE HAD __Grey__ EYES,

__Brown__ HAIR, __Ruddy__ COMPLEXION, AND WAS __Five__ FEET

__0__ INCHES IN HEIGHT. HIS COLOR WAS __White__

GIVEN UNDER MY HAND AT __Camp DG-4G-A__, THIS __Twelfth__ DAY

OF __July__, ONE THOUSAND NINE HUNDRED AND __Thirtyseven__

C.C.C. Form No. 2
April 5, 1933

* Insert name, as "John J. Doe".
** Give reason for discharge.

3—10171

JAMES O. YOUNTS, 1st Lt., Inf-Res.,
(Name)   (Title)
Commanding.

Courtesy of the Mohave Museum of History and Arts, Kingman, Arizona.

The location of a CCC camp made no difference as far as the daily routine was concerned. Each day started at 6:00 a.m. and by 6:30 a.m., in their working clothes, the men took their physical training excercise, followed by breakfast. After breakfast there was time to clean the quarters, attend roll call, and inspection. At 7:45 a.m. the men walked or rode to work depending on the distance. Work continued until noon, followed by lunch which was usually delivered to the work site. The lunch break lasted for an hour, and work continued until 4:00 p.m., after which time the men returned to their camp.

RECORD OF SERVICE IN CIVILIAN CONSERVATION CORPS

** Served:

a. From __4/22/36__ to __9/30/36__, under __Interior__ Dept. at __Co. 2865, Camp DG-46-A,__
Kingman, Arizona.
Type of work __Road Construction__ *Manner of performance __Excellent__ _Jay_

b. From __10/1/36__ to __3/31/37__, under __Interior__ Dept. at __Co. 2865, Camp DG-46-A,__
Kingman, Arizona.
Type of work __Carpenter__ *Manner of performance __Excellent__ _Jay_

c. From __4/1/37__ to __7/12/37__, under __Interior__ Dept. at __Co. 2865, Camp DG-46-A,__
Kingman, Arizona.
Type of work _Carpenter_ *Manner of performance _Excellent_ _Jay_

d. From _____ to _____, under _____ Dept. at _____
Type of work _____ *Manner of performance _____

e. From _____ to _____, under _____ Dept. at _____
Type of work _____ *Manner of performance _____

Remarks: __Last discharged from Co. 2865. Serial No. CC9-2865190.__
The Project Superintendent makes the following estimate of the individual: The
services of Enrollee McB███ were performed in a satisfactory manner.

____Enr; Add: Box 905, Kingman, Arizona.____

JUL 24 1937

FORT BLISS, TEXAS
PAID IN FULL $.6.00
CARL HALLA, LT. COL., FD

Discharged: __July 12, 1937__ at __Camp DG-46-A, Kingman, Arizona.__

Transportation furnished from __Camp DG-46-A__ to __Kingman, Arizona.__

(Name) (Title)
JAMES O. YOUTS, 1st Lt., Inf-Res., Commanding

* Use words "Excellent", "Satisfactory", or "Unsatisfactory".
** To be taken from C.C.C. Form No. 1.

U. S. GOVERNMENT PRINTING OFFICE 1935   2—10171

Courtesy of the Mohave Museum of History and Arts, Kingman, Arizona.

The evening meal was served at 5:30 p.m., and the time between the return from work and the meal was free. Some men used this time to participate in sports or in the library. After the evening meal many of the men pursued their education by participating in the camp program of study. Others tried to occupy themselves playing table tennis or even by going to town if the camp was nearby. They had to be back by "lights out" at 9:45 p.m. Lights were flashed off for the preparation of the night's rest, and lights went out at 10:00 p.m. At 11:00 p.m. a camp official checked to see that all men were present.

The men in each camp could see their families once a month if they so desired, depending upon the distance. Two hundred miles was the limit for visiting home. At the time of reenlistment there was a six-day leave with full pay. In addition to normal leave privileges, all official holidays were observed with no work, including religious holidays. All men of voting age were given three days with pay to register and vote in primaries, and local, state, and national elections.

The CCC men were also free on week-ends. Saturdays were usually devoted to sports and group activities such as drama or choral work. On Sundays religious services were held in all camps.

Each camp had a baseball and football team. They played games against other camps and local high school teams. Boxing was a popular sport in the camps and some very good boxers developed in matches in Kingman and Needles.

There were eight camps in Mohave County, including Hualapai Mountain Camp (forest); Round Valley Camp (soil erosion, highway construction and repairs); Francis Creek Camp (soil erosion and road building); Buck and Doe Camp, north of Peach Springs; Hualapai Valley Camp, near Red Lake. McConico Camp (soil erosion) was in Coconino County, and there were two camps in the Grand Canyon area. There were probably other camps in the Flagstaff area and near towns along Route 66.

A Civilian Conservation Corps crew loading poles for telephone lines, Hualapai Mountain Park project near Kingman, Arizona, 1936.
Courtesy of the Mohave Museum of History and Arts, Kingman, Arizona.

PART III:
PLACES
ON THE MAP

PLACES ON THE MAP

This book would not be complete without "Places on the Map." Included are the cities of Flagstaff and Kingman; many small towns that owe their beginning to the Atlantic & Pacific Railroad; the early mining camps of Oatman and Chloride which have survived as tourist stops; other mining camps which have vanished with the ages; some communities first settled by members of the Church of Jesus Christ of Latter Day Saints (Mormons); Hoover, Glen Canyon and Davis dams; five national monuments; Kanab (Utah), Needles (California) and Laughlin (Nevada), all just across the Colorado River and important because their histories are intertwined with that of Northwest Arizona.

The history of man in Northwest Arizona is seen in the places where he traveled and lived, the hardships and challenges he faced, his triumphs and failures. The best of man's achievements in Northwest Arizona are documented in this chapter, which is geographically separated into three sections: the Western Area, South of the Grand Canyon and the Colorado River, and North of the Grand Canyon and the Colorado River, (see map, pages 142 and 143).

Peach Springs Trading Post located along the National Old Trails Highway, 1922. Courtesy of the Mohave Museum of History and Arts, Kingman, Arizona.

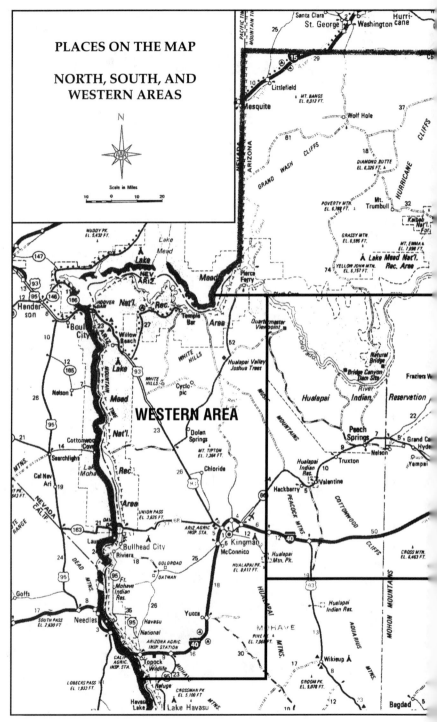

PLACES ON THE MAP

NORTH, SOUTH, AND
WESTERN AREAS

N

Scale in Miles

Modified from AAA Road Atlas, 1989.

142

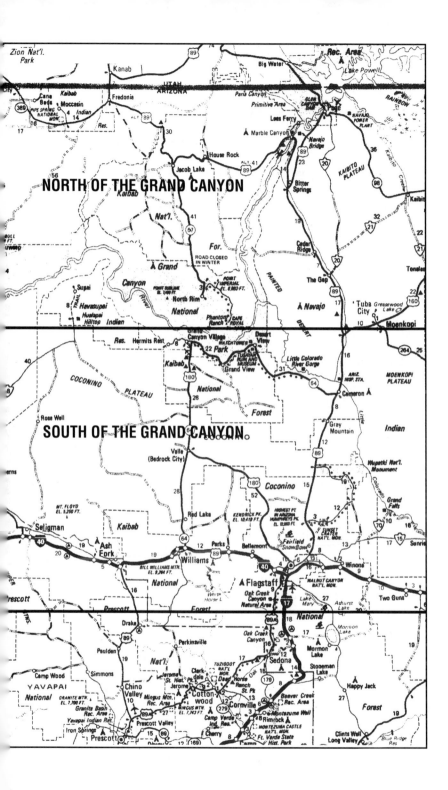

NORTH OF THE GRAND CANYON

SOUTH OF THE GRAND CANYON

THE WESTERN AREA

Bullhead City (El. 540')

Bullhead City covers the area that once was Hardyville, and originated as a private development built for workers on the Davis Dam project in 1945. The town's name is derived from Bullhead Rock which is now largely concealed by the waters of Lake Mohave.

In the 1960s there was a marketing push in southern California for people to buy inexpensive mobile homes and park them close to the Colorado River. A second growth wave came in the late 1970s and early 1980s as the casinos expanded in Laughlin across the river.

In 1980 the population of Bullhead City was 10,000. In 1987 the Laughlin Bridge was completed, providing access between Laughlin and Bullhead City without having to travel across Davis Dam.

By 1990 the population was 25,000. Today the area is a popular retirement and year-round recreation community of 34,000 and still growing.

A new subdivision of 40,000 upper-end homes is now being constructed in Bullhead City, and there is no end in sight.

Camp Beale Springs — Site (El. 4,000')

Two springs are located in a shallow drainage about four miles northwest of Kingman. They were long known and used by Native Americans along the east-west Rio Grande - Pacific Ocean trail.

Lt. Edward F. Beale camped at the springs on his 1857 survey for a wagon road along the 35th Parallel. A few years later a mail station was established and there were continual skirmishes with the Haualapai Indians.

In 1867 the mail station became known as Beale Station and later that year the name was changed to Beale Springs.

In the spring of 1871 a small military tent camp was established at the springs by Company F, 12th U.S. Infantry out of Fort Whipple for the purpose of protecting travelers along the Prescott and Mojave Toll Road. In 1873 a post office was opened, and at the same time an Indian Agency office, which would be used as a feeding and supply station for the Hualapai, was established. The post was abandoned in 1874, became a campsite and way station along the toll road, and remained active until the National Old Trails Highway was designated in 1914, passing through Kingman.

In 1883 Beale Springs was the source of water for Kingman. Water was first hauled by wagon and then piped into Kingman from a small reservoir that remains as an empty basin today.

Little can be seen today of Camp Beale Springs except some rock walls and foundations.

Chloride (El. 4,009')

Located about 20 miles north of Kingman along the western side of the Cerbat Mountains, the mining camp of Chloride was established in 1863 after silver was discovered in the area. Named for the silver chloride mineral

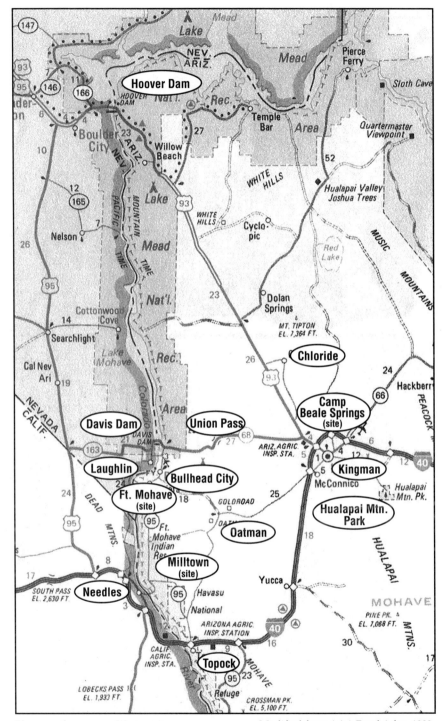

Places on the map — Western Area.   Modified from AAA Road Atlas, 1989.

cerargyrite, the camp grew slowly because of hostilities with the Hualapai Indians. After a peace treaty in 1870, about 1,000 prospectors rushed to the area, and by 1872 the camp had a brewery, general store, blacksmith shop and several saloons. Lumber sold for $75 to $100 a 1,000, canned goods were $1 a can, sugar and coffee were $1 a pound, and miners wages were $5 a day.

Chloride boomed in the 1890s and by the turn of the century had a population of 1,500. The town was serviced by stagecoach and the Arizona & Utah Railroad, a branch line of the Santa Fe Railroad from Kingman, built in 1899. Chloride, with a population of 2,000, became the distribution center for nearby camps, mines and other districts.

By 1904 mine production slowed, and by 1910 the population had dwindled to a few hundred, lasting through World Wars I and II.

Many movies and commercials have been filmed in and around Chloride throughout the years.

Among the ruins of mines and mills, the population in 1990 was about 300. Today Chloride is the home of artists, writers, musicians and craftsmen. About 20,000 tourists visit Chloride each year.

Davis Camp — Site (El. 585')

Davis Camp was located about three miles north of Bullhead City near Davis Dam. It was the former federal housing area for government workers who built the dam between 1942 and 1950 and later maintained it. It grew to over 100 homes and boasted a grocery store, gas station, post office, recreation center, swimming pool, two churches and various service buildings.

Directly across the river in Nevada was another small community that housed some of the contractors. In 1982 the Bureau of Reclamation declared the Davis Camp area surplus, and it was turned over to Mohave County to develop into a regional county park.

Davis Dam (El. 661')

Davis Dam is located a few miles upriver (north) from Bullhead City, Arizona and Laughlin, Nevada. The dam and powerplant facility were constructed by the Bureau of Reclamation in Pyramid Canyon.

As early as 1902 the Reclamation Service investigated a possible dam site in Pyramid Canyon. It was originally called Bullhead Dam after a rock formation near the proposed site. It wasn't until 1941 that the project was authorized and the name changed to Davis Dam in honor of Arthur Powell Davis, Director of Reclamation from 1914 to 1923.

In 1942 construction began on Davis Dam and steamboat traffic up the Colorado River ceased. Because of the onset of World War II construction was halted after a few months.

The housing area for the construction workers on the dam was named Bullhead City. The government employees lived in Davis Camp.

In 1946 after World War II, construction resumed, and in January 1950 the first water was impounded forming Lake Mohave, 160 feet deep and more than 30 miles long. The dam, completed in 1953 at a cost of $67 million, is an earth and rock-filled embankment with a concrete spillway and powerplant.

Davis Dam is part of the Lower Colorado Dams Project which includes Parker Dam. Parker and Davis dams operate in unison with Hoover Dam to control floods along the river and furnish hydroelectric energy to the Southwest. In accordance with a 1944 water treaty with Mexico, river flow at Davis Dam is regulated so that 1.5 million acre-feet of water can be delivered to Mexico annually.

The Davis Dam Powerplant generates one to two billion kilowatt-hours of electricity annually which is used in the Southwest.

Fort Mohave and Mohave City — Sites (El. 541')

Fort Mohave and Mohave City were located eight miles south of Bullhead City.

In January 1859 Colonel William Hoffman was ordered to find a location for a fort to be used to protect the Colorado River crossing on Beale's Wagon Road from warring Mojave Indians (a branch of the Yuma Indian Tribe).

Camp Colorado was established at Beale's Crossing on the Arizona side of the Colorado River, and Major Lewis Addison, who was placed in charge of the fort, renamed it Fort Mohave on May 1, 1859.

The post was abandoned in 1861 at the outbreak of the Civil War and re-established in May 1863.

Mohave City was founded in early 1864 just beyond the northern limit of Fort Mohave military reservation. By August there were two saloons, three stores and new adobe buildings going up. The town was a "recreation" town for Fort Mohave, catering to the needs of the soldiers. The saloons were the scene of drunken brawls and at least a dozen killings.

That same year ferry boat service across the Colorado River began.

In October 1864 the Territorial Legislature selected Mohave City as the first seat of Mohave County, which it held until 1867.

In 1866 a post office opened.

In 1869 Fort Mohave expanded its military reservation boundaries and absorbed Mohave City. Many years of litigation followed between Mohave City residents and the federal government.

In 1890 the old fort and 14,000 acres were transferred from the War Department to the Department of the Interior and became an Indian residential boarding school. The school closed in 1931.

On February 2, 1911, by Executive Order, the Fort Mohave military reserve, including Mohave City, was included within the Mojave Indian Reservation.

Hardyville (El. 535') (See Bullhead City)

In 1864 Captain William H. Hardy established a small settlement along the Colorado River some five miles above Fort Mohave. Named Hardyville, it became a shipping point for goods brought by steamboat up the river, to be shipped overland to the mines in the Cerbat Mountains and elsewhere.

Hardy built a ferry across the Colorado River along the new Prescott and Mojave Toll Road, which became the main wagon road between Prescott, Arizona and Los Angeles, California.

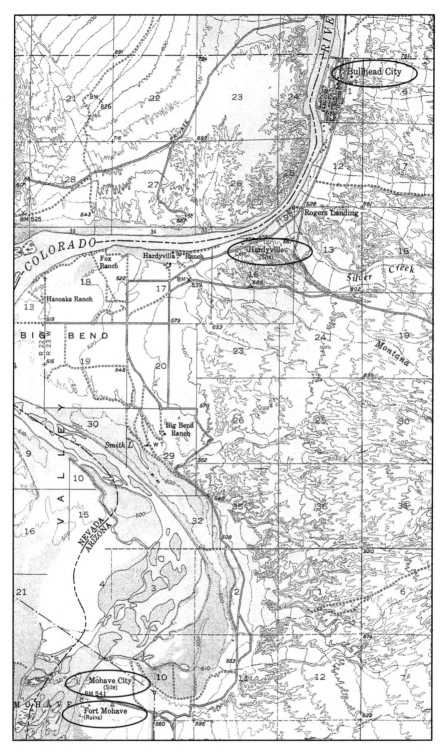

Sites of Hardyville, Mohave City and Fort Mohave.
U.S.G.S. 15' Davis Dam, Nev.-Ariz.-Calif. Sheet, 1952.

A post office was established in 1865.

From 1867 to April of 1873 Hardyville served as the second county seat of Mohave County, after which it was moved to Cerbat.

In 1870 Hardyville had a population of 20. In November 1872 the town was destroyed by fire and again in 1873. The post office closed in 1883. Hardyville was rebuilt in 1906 as the closest supply point to the Oatman area. Few traces remained when Bullhead City was established in 1945.

Hoover Dam (El. 640' at river level and 1,220' at full Lake Mead)

After the Colorado River flood of 1905, which created the Salton Sea, a solution was needed for the Colorado River flood problems. The answer was the construction of several dams across the river. Congress authorized the Boulder Canyon Project in 1928 to control flooding, to improve river navigation, to provide for storage and delivery of water, and to generate electrical energy. The site selected was Black Canyon about 25 miles southeast of Las Vegas, Nevada.

The Bureau of Reclamation designed the dam and supervised its construction, contracting with six of the largest dam builders in the country.

Preliminary work building a construction camp that would house some 3,500 workers (now called Boulder City, Nevada) began in 1931. This also included funding and developing large sources of sand and gravel to be used in the production of concrete; construction of access roads, a concrete batch plant, a coffer dam and diversion tunnels; drilling and blasting rock

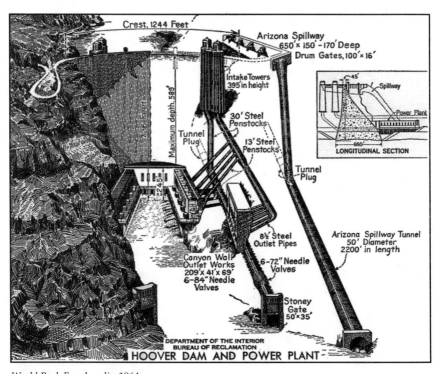

World Book Encyclopedia, 1964.

foundation work; the erection of all the necessary equipment required for the project; and construction of electric power lines.

Construction began in 1932 when coffer dams blocked the river water, forcing it to flow through the diversion tunnels.

On June 6, 1933 the first concrete was poured into forms 25 to 60 feet deep. Cableways were used to transport buckets of concrete 24/7 at 1,100 buckets a day. More than 4.4 million cubic yards (six million tons) of concrete were poured. Concrete pouring was completed in the summer of 1935, and the dam was completed two years ahead of schedule and under budget. In September 1935 President Franklin Delano Roosevelt dedicated the dam.

The completed arch-gravity dam, the highest in the United States, is 726 feet high, 1,244 feet long, with a base of 660 feet. There are four water intake towers built in front of the dam that provide water through steel-lined penstocks to turn its 17 turbines, each 30 feet in diameter. Four to five billion kilowatt-hours of electricity are produced annually.

Spillways, 50 feet in diameter, have been used only once (in 1983) to avoid flooding.

The dam cost $120 million and the entire project about $385 million.

The project provided work for more than 5,000 men during the deepest part of the Great Depression. The project was also the first where hardhats, homemade by dropping two baseball hats into hot tar and then placing them together until they hardened, were used.

In 1931 the dam was unofficially called Hoover Dam in honor of President Herbert Hoover. The name was changed to Boulder Dam in 1933, and in 1947 Congress officially renamed it Hoover Dam.

Lake Mead is impounded by Hoover Dam and is the largest man-made lakes in the United States by volume of water. It took six years to fill and

Hoover Dam, March 2005 (note the low water level of Lake Mead).   Photo by L.J. Ettinger.

when full has a 550-mile shoreline, is 110 miles long, 500 feet deep, and averages 200 feet in depth with a storage capacity of 10 trillion gallons (32.3 million acre-feet) of water. The weight of the water in Lake Mead has caused a number of earthquakes felt in Boulder City, the largest of which was 5.0 on the Richter Scale.

When Lake Mead was full water flowed into the intake towers and then into turbines which turned generators, producing electricity. By 1987 enough power was generated to pay for the dam with interest.

In 1964 the Lake Mead National Recreation Area was established as the nation's first national recreation area. It extended about 140 miles along the Colorado River from the western boundary of Grand Canyon National Park to Bullhead City, Arizona, and embraced 1.5 million acres in northwest Arizona and southern Nevada. Today there are six major recreational centers with marinas and launch facilities which are visited by more than nine million people each year.

Called one of the Seven Engineering Wonders of the World, each year more than one million visitors visit Hoover Dam, and more than eight million enjoy recreation activities at Lake Mead.

For security reasons the events of September 11, 2001 changed the traffic patterns at the dam. No large trucks are allowed to drive across the dam, causing a detour of about 95 miles. A new bridge is under construction across the Colorado River just downstream from the dam and, when completed in 2008, through traffic will no longer cross the dam.

Hualapai Mountain Park (El. 6,700')

Hualapai Mountain Park is located 14 miles south of Kingman. It was originally a primitive mountain getaway with a campground next to a Boy Scout camp which was reached by a rough dirt road.

In 1934 a Civilian Conservation Corps (CCC) camp for 200 men was built at the base of the Hualapai Mountains south of Kingman. Their project was to build the Hualapai Mountain Road and improve an existing connecting road. In September 1934 at the summit, an area was cleared of growth, trails were constructed, and two recreation buildings were started.

In June 1935 a new CCC work camp was opened in the Hualapai Mountain Park. The original road to the park was a steep, narrow and dangerous mountain trail. The new road to Kingman was a highway six miles shorter, and could be traveled by all vehicles of that time. New roads were also built to other scenic parts of the Hualapai Range. Constructed in the park were four large cabins, water lines to the new camping and cabin area, restrooms, garbage pits, cattle guards and fences.

Kingman residents found new picnic and recreational facilities next to the CCC camp. The camp mess hall, built of logs, became the Hualapai Mountain Lodge.

The CCC camp was closed in 1937, leaving a beautiful park which could be reached in a half hour from Kingman.

Abundant wildlife includes mule deer, elk, mountain lion, fox and a variety of birds.

KINGMAN (El. 3,336')

Kingman, named after Lewis Kingman who supervised construction of the Atlantic & Pacific Railroad, was first established in 1882 as the Kingman Siding of the Middleton Division of the A&P, a watering stop along the line. Water was obtained from wells drilled by the A&P near the depot.

In 1883 Kingman had a post office and its tax rolls included one tent used as a residence, a lumber house, saloon, hotel, stable and corral with one horse. Water for the town was hauled in by wagon from Beale Springs.

The Kingman Sampling Works was established in 1883 where small-scale mine operators could bring their ore to be assayed and processed. The sampling works operated until January 2, 1901 when it was destroyed by fire.

Kingman slowly grew, drawing some of the established businesses away from Mineral Park. In 1886 *The Mohave County Miner,* a newspaper established at Mineral Park in 1882, relocated to Kingman.

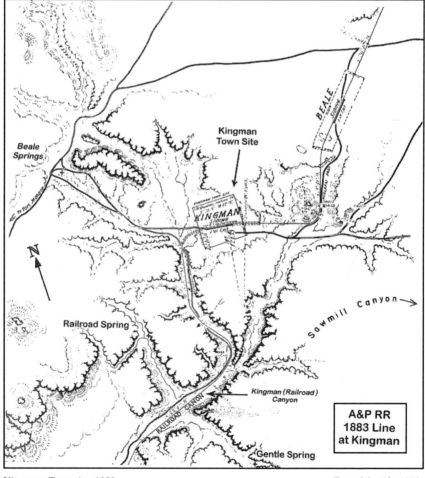

Kingman Townsite, 1883.   From Myrick, 1998.

The election of 1887 gave the Mohave County seat to Kingman. By 1888 most of the original townsite lots were sold near the railroad terminal. On June 17 Kingman suffered its first major fire that destroyed an entire block of buildings.

By the end of the 1880s Kingman rivaled Mineral Park in size with a population of 300 in 1890. Each residential property had its own water well and storage tank.

July 4th was the big event of each year, celebrated with Indians, orators, dances, a rodeo and a miner's drilling contest. At other times circuses and wild west shows arrived in town on the railroad and played to large and enthusiastic audiences.

On May 16, 1892 Kingman's second major fire consumed 26 buildings.

In 1899 the Arizona & Utah Railroad completed its line between Kingman and Chloride, making the shipping of supplies and ore concentrates easier and cheaper. Kingman's telephone directory listed 23 names.

The discovery of the Gold Road Vein 30 miles southwest of Kingman in May 1900 helped establish Kingman as a center for mining activities in Mohave County. The official Kingman Township Town Plat was completed showing streets and alleys. The population was approximately 550.

Also in 1900, on April 28, the Kingman passenger and freight station burned.

In 1901 the Fred Harvey Company opened a small lunch room near the Santa Fe Depot.

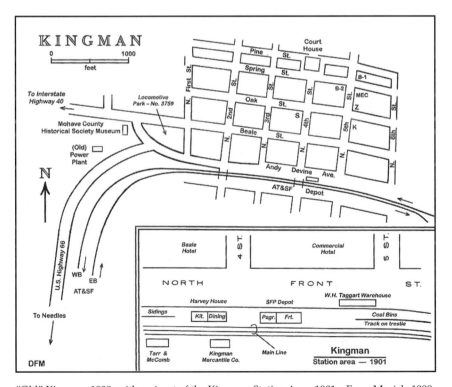

"Old" Kingman 1998, with an inset of the Kingman Station Area, 1901. From Myrick, 1998.

153

In 1902 the Oatman and Gold Road area was active, and the economic impact on Kingman showed with an increase of 650 residents by 1906.

In 1906 Tom Devine purchased the Beale Hotel for $32,000. His son, Andy, would become a movie and television star in later years.

On June 6 another fire consumed several buildings on North Front Street and a month later the Santa Fe Depot burned. The city fathers now made a serious attempt to organize a fire department.

The Desert Power and Water Company's power house became operational in 1909 supplying power to area mines, Kingman homes and businesses. The same year, the three-story Hotel Brunswick opened.

Kingman's first service station: Roy's Service Station and Free Campgrounds, along National Old Trails Highway on South Front Street (now Topeka). "Roy's buys and sells new and used tires and cars. Second-hand goods of all kinds. Trades on Everything." 1910.
Courtesy of the Mohave Museum of History and Arts, Kingman, Arizona.

In 1912 Arizona became the 48th state. Kingman had a population of 1,900 and mining was still the principle industry in the area. The National Old Trails Highway came through Kingman bringing more automobile traffic.

The new Mohave County Courthouse was dedicated and opened in 1915.

A building boom ensued, and in 1919 Kingman had seven garages, three meat markets, two drug stores, a picture show and numerous hotels and saloons. Kingman's only industry was the Yucca Fiber Factory which made rope from yucca plants.

In 1926 the National Old Trails Highway was designated as U.S. Route 66, which was the main street through Kingman.

In 1928 Kingman Municipal Airport opened. In 1929 the first Ford Tri-Motor "Tin Goose" airplane landed at "Port Kingman," the first airport in Arizona, laid out by Charles Lindbergh as the first stop on his Transcontinental Air Transport Company (T.A.T.), a forerunner of TWA.

In 1930 there were six traffic stop signs at intersections in Kingman. The first all-talking movie came to town that same year.

Between 1934 and 1942 two CCC camps were built near Kingman, each housing 200 young men who worked on nearby projects.

Kingman continued to grow with mining in the Oatman District and the construction of Hoover Dam and Route 66. With completion of Hoover Dam in 1938 and the generation of inexpensive electric power, the Kingman Powerhouse could not compete and closed.

Kingman, Arizona, pre-1947.
Courtesy of the Mohave Museum of History and Arts, Kingman, Arizona.

In 1942 the Army Air Force opened the Kingman Army Airfield which served as a Flexible Gunnery School to train aerial gunners. With an Army population of 17,000 officers and men, an estimated 36,000 gunners were trained here.

After World War II ended, the Kingman Army Airfield became a storage depot for over 7,000 planes, including B-24 Liberators, B-17 Flying Fortresses, B-29 Super Fortresses, B-25 Mitchell bombers, P-38 Lightnings, P-47 Thunderbolts, P-40 Flying Tiger fighters, and A-26 medium bombers. Most were sold for scrap.
Courtesy of the Mohave Museum of History and Arts, Kingman, Arizona.

Kingman was incorporated on January 21, 1952 and became a major crossroad for travelers going east and west on Route 66 and north and south on U.S. 93.

In 1968 the Mohave Museum of History and Arts was dedicated by Secretary of the Interior, Stewart Udall. In 1971 Mohave Community College opened in Kingman. In 1978 Kingman declared Andy Devine Days to be an annual event. In 1980 U.S. Interstate 40 opened, by-passing downtown Kingman, and in 1997 the renovated Hotel Brunswick and Power House opened.

Because of its high-desert climate, Kingman and surrounding area are rapidly developing as a prime location for retirees and those who provide services for them.

Today Kingman has a population of more than 25,000.

Andy Devine: Andy was born on October 7, 1905 in Flagstaff, Arizona. His family moved to Kingman in 1906 and his father purchased the Hotel Beale. Andy attended school and played football and basketball for Kingman High School. After graduating high school, he attended Santa Clara University in California where he lettered in football. In 1925 Andy played professional football in Los Angeles.

Andy first appeared on the movie screen in two-reel silent movie comedies in 1926. His raspy voice was his signature, the result of a childhood accident when he fell on a curtain rod and damaged his larynx.

While he was developing his acting skills, he held jobs as a telephone lineman, lifeguard and a news photographer.

In 1933 he met actress Dorothy House, who became his wife and bore them two sons, Tad and Dennis.

He played the stagecoach driver in the 1939 classic movie "Stagecoach." Andy was a regular on the Jack Benny radio show. He appeared in over

Andy Devine (L) and Guy Madison from the TV Series "Wild Bill Hickock," 1950-1956.
Courtesy of the Mohave Museum of History and Arts, Kingman, Arizona.

400 movies and, for six years played Jingles in the early television series "Wild Bill Hickock."

As Kingman's favorite son, a street in Kingman was named after him in 1955, featured on the television show "This is Your Life."

Andy returned to Kingman often, appearing in Andy Devine Days parades.

In 1957 he was diagnosed with diabetes and in 1973, leukemia. Andy passed away on February 18, 1977 of cardiac arrest at the age of 71. His wife, Dorothy, died on June 14, 2000 at age 85.

Laughlin, Nevada (El. 535')

In 1953 Davis Dam was completed, and two small clubs and some fishing camps were opened on the Nevada side of the Colorado River, called "South Pointe."

In 1959 the Fort Mohave Valley Development Law was enacted whereby thousands of acres of land along the Colorado River would be administered by the Colorado River Commission.

In 1966 Don Laughlin bought the Riverside Bait Shop and re-opened it as the Riverside Resort. Also in 1966 the federal government turned over the first 4,000 of 15,000 acres to the state of Nevada. Some of the land was sold to developers, and four square miles to the south were sold to Southern California Edison for construction of a coal-fired electricity-generating plant.

A year later Oddie Lopp purchased the Bobcat Club, and Southern California Edison Company began construction of their plant that would utilize water from the Colorado River for cooling. Coal from the Four Corners Area of Arizona would be piped in a slurry pipeline almost three hundred miles.

Parking lots were built on the Arizona side of the Colorado River and shuttle boats were used to carry passengers to the Riverside Resort and Bobcat Club. In 1968 a third casino, the Monte Carlo, opened and the U.S. Postal Inspector officially named the community Laughlin.

During the 1970s the three casinos prospered and expanded. By 1972 there were 30 condominiums, 10 private homes and a trailer park in Laughlin with a population that had grown to about 100.

The beginning of new casino construction in 1977 attracted big-time investors such as Circus-Circus (Edgewater Hotel and Colorado Belle), 1982; Sam's Town (Gold River-River Palms), 1984; Ramada (Ramada Express), 1988; Harrah's (Harrah's Laughlin), 1988; and Flamingo Hilton, 1990.

In 1987 the Laughlin Bridge opened which linked Laughlin with Bullhead City, Arizona. Since 1988 the major casinos have more than doubled their room capacity.

In 1990 Laughlin's population jumped to 4,800, and in 1991 Laughlin High School opened along with a new airport facility to accommodate commercial flights. The first jet would land in 1993.

In 1997 the Horizon Outlet Mall opened with 52 retail stores and a movie theater complex. In 1998 the Gold River became the River Palms Resort.

Today, Laughlin has a population of more than 7,000.

Milltown — Site (El. 800')

Milltown was located in the S1/2 Section 21, T.18N., R.21E.

The Mohave Gold Mining Company purchased the Leland Mine in the Oatman Mining District in 1903. A 40-stamp mill was constructed 11 miles southeast of the mine even before ore reserves were found. Located west of the mountains along the gentle slope to the Colorado River, the mill workings were called Milltown.

A pumping plant on the Colorado River two miles below Needles pumped water about six miles to the mill.

A 17-mile narrow-gauge railroad was built from the mine to the Colorado River at Needles, Arizona servicing Milltown. A ferry crossed the river and a connection was made with the Santa Fe Railroad.

Only about 4,500 tons of gold ore was treated before severe flooding in 1904 washed out all but two miles of the tracks between the mill and the river. Without the railroad, the mine and mill shut down.

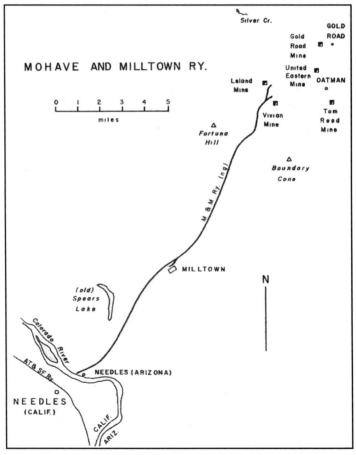

Mohave and Milltown Railway.   From Myrick, 1963.

Needles, Arizona (El. 480')

Named after the sharply pointed peaks below Red Rock, Needles was a small settlement along the Colorado River that was abandoned when the Southern Pacific Railroad reached Needles, California.

In 1903 Needles, Arizona revived as the Colorado River terminus of the Mohave and Milltown Railroad, which was constructed to bring ore from the mines in the Oatman Mining District to Milltown for processing and then on to the Needles ferry crossing and steamer landing. There were storage yards, machine shops, several businesses, and houses were going up as fast as men and lumber were available. The community was abandoned after the flood of 1905.

NEEDLES, CALIFORNIA (El. 488')

Named for a group of three sharp peaks surrounded by low hills below Red Rock, Needles was a small tent town built for the arrival of the railroad in late 1882.

On May 19, 1883 construction crews laying the Southern Pacific tracks eastward from Mojave, California arrived at Needles. Atlantic & Pacific tracks westward across Arizona arrived at the Colorado River near Needles on May 29, 1883. Needles had a construction population of about 600, but quickly became a ghost town for a short period as the crews were sent back to San Francisco. Maintenance crews were quickly brought in and businessmen and their families soon arrived.

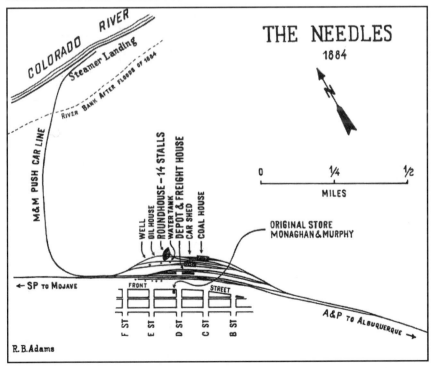

The Needles, California, 1884.   From Myrick, 1963.

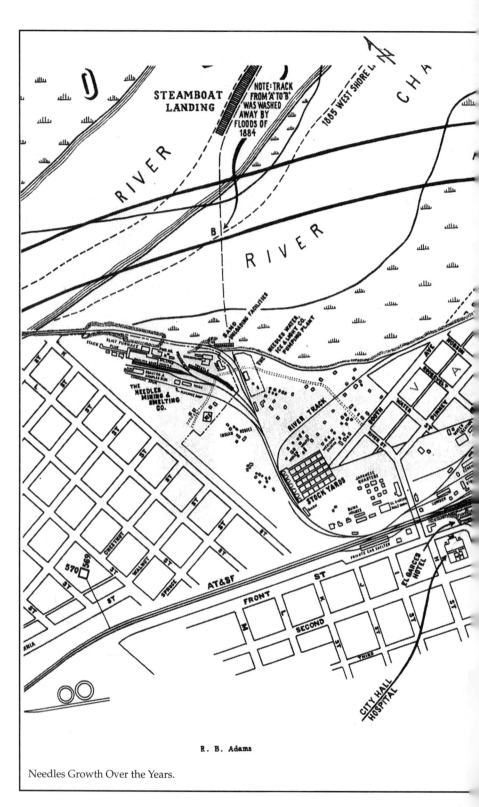

Needles Growth Over the Years.

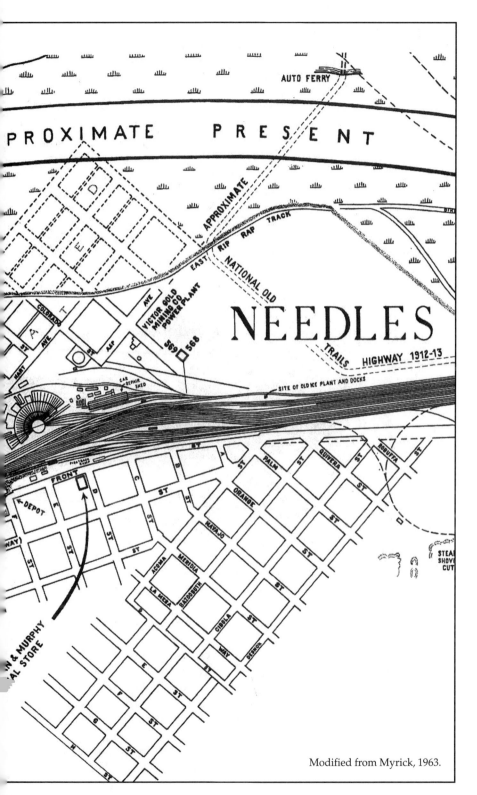

Modified from Myrick, 1963.

The Southern Pacific built a magnificent depot capable of accommodating a hotel, car sheds, shops and a 14-stall roundhouse. Within a month Needles boasted a Chinese wash house, newsstand, restaurant, several grocery stores and nine or 10 saloons.

With much resistance from the Colorado River at flood stage, a "flimsy looking bridge," 1,600 feet long, was completed by the A&P with difficulty. The bridge was an obstruction to navigation because it lacked a draw.

In 1884 the A&P leased the SP's route from Needles to Mojave, California.

Colorado floods swept the bridge away in 1884 and also its two successors in 1886 and 1888. Finally the A&P designed a high cantilever span which was completed at a narrower point 10 miles downstream in May 1890.

Citizens of Needles enjoyed cultural activities. There was a literary society, a theatrical group, and many musicians, including the Needles Brass Band.

In the early 1890s Needles was the supply center for new mines upriver at the camps of Eldorado Canyon and Pyramid. Eastern capitalists proposed to make Needles a milling and smelting center.

On September 19, 1891 a fire swept through the business section of town that had fingers pointing to Chinese opium dens.

In 1892 the Needles Reduction Company was organized, and a cyanide works and smelting furnace were erected north of town. The mill began operation in April 1892, treating gold ore until March 1894 when operations ceased.

Needles was extremely hot in the summer and could be miserably cold in the winter. The A&P maintained a large storage house to accommodate dozens of carloads of imported ice. During the summer months the railroad re-iced cars of perishable fruits and vegetables on eastbound trains.

In 1893 the Santa Fe took over the A&P operations through Needles.

In early 1894 a 30-ton ice manufacturing plant was erected. Most of its ice was for east-bound California fruits and vegetables. Ice was also sent for storage at Mojave and Barstow, California and Bagdad and Peach Springs, Arizona.

In 1897 the Santa Fe took over the A&P operation.

The Needles Smelting Company was organized in August 1900, and their smelter began operation in December of that year. Initially treating lead-zinc ore, the smelter quickly switched to treating copper ore, which removed most Wallapai District mines from its list of customers. The smelter burned in August 1904.

The Santa Fe Railroad erected a recreation building in Needles in late 1902 which offered its employees a place to spend time on layovers.

In April 1905 the Needles smelter moved a mile upriver and rebuilt, treating gold, lead-silver-zinc and copper ores at 150 tons per day.

Like many early western towns, Needles suffered its share of fires. On September 9, 1906 the old depot-hotel burned with the loss of several lives.

A new Harvey House hotel, called El Garcés, was built in 1908 and closed in 1949.

In 1909 the United States Smelting, Refining, and Mining Company purchased the Needles smelter and doubled the capacity to 300 tons per day.

Ferrying across the Colorado River at Needles, California, pre-1916.
Courtesy of the Mohave Museum of History and Arts, Kingman, Arizona.

Mine operators chose to send their zinc ore to Oklahoma for treatment in 1912 and the Needles Smelter closed and finally shut down in 1917.

At first most people traveled to and from Needles by rail or steamer. The wagon road passing through Needles became the Old Trails Highway in 1914, and gave access to early automobile travel across the country. Crossing the Colorado River was by ferry until 1916 when the first automobile bridge was constructed at Topock. The Old Trails Highway was designated as U.S. Highway 66 (Route 66) in 1926 and the road was greatly improved.

Construction crews on the Davis and Parker dam projects used Needles as a materials distribution point.

During World War II many military trains and convoys passed through Needles. Areas near Needles were the site of military training maneuvers by General Patton's army.

Dredging of the Colorado River by the Bureau of Reclamation in the early 1950s ended the long history of flooding each spring. The dredging dammed the valley floodplain lands near Needles, making them suitable for agriculture and housing. The water of the Colorado became clear and suitable for recreation.

Local citizens lobbied for years for Interstate 40 to pass through Needles, which it did and this greatly helped the local economy.

In the early 1960s Needles population was about 4,500, and today the population of Needles is near 5,000.

OATMAN (El. 2,700')

Before Oatman

Rich discoveries of gold at Gold Road in 1900, Snowball in 1901, and along the Tom Reed Vein in 1904 created a small boom and brought several hundred people to the area who resided in the camps of Gold Road, Snowball and Vivian (see map, page 55).

Within an area of several miles surrounding the eventual location of Oatman, a number of mining camps would be established as new mines were discovered. There were no roads into or out of the area, with the exception of Beale's Wagon Road which went over Sitgreaves Pass and then down Silver Creek to Hardyville.

Nearly every mine had its own camp where miners and employees lived, mostly in tent houses, and the camp usually survived only as long as the mine was operating. When the mine shut down the miners would leave, looking for new jobs at other mines.

After the discovery of gold about 1.5 miles north of present-day Oatman in 1901, Gold Road Camp was established. A post office opened in 1902. By 1907 Gold Road was a camp of a few wooden stores and 300 to 400 tent houses. The camp declined from 1907 to 1911, prospered during the Oatman boom, and declined again during the Great Depression of the 1930s. The post office closed in 1942.

Snowball, established in 1902 about one mile southwest of present-day Oatman, probably survived until 1905. The camp had a Chinese restaurant, two saloons and a number of family tent houses. Water was packed on burros from the Treadwell Mine shaft to the camp. A miner's union was organized, officially known as the Snowball Union.

Vivian was constructed in 1904, about one-fourth mile west of Snowball, and lasted until 1909. There were stores, saloons and a post office which was in service from 1904 to 1909.

The Tom Reed Mine was taken over by the Blue Ridge Gold Mine Company of Pasadena, California in 1905 and the camp of Blue Ridge was founded. The company explored and developed the property, built a 10-stamp mill, and went into production. The venture was not successful and the mine was sold to the Tom Reed Gold Mining Company in 1906 who operated it until 1932. Schrader's map of 1909 (see page 55) shows the Blue Ridge Camp about one-fourth mile northwest of the site of future Oatman.

Oatman Camp

The name Oatman first appeared in the "Mining News" in 1912 (Ransome, 1923). The camp was in a narrow gulch called Boundary Cone Wash and was dependent upon the Tom Reed Mine. It is likely that the camp of Oatman began sometime between 1909 and 1912 and that Blue Ridge Camp quickly gave way to the camp of Oatman.

The name Oatman has several possible origins. One is that it was named for Olive Oatman who, with her family, was captured by Indians in 1851 and was later rescued in 1857 near the present site of Oatman. Another is that the camp was named for John Oatman, a local miner who claimed to be the son of Olive Oatman and a Mojave Indian.

There was continued exploration along the Tom Reed Vein and in 1914 the United Eastern Mining Company discovered buried gold deposits along the vein.

This was also the year that the wagon road traversing Boundary Cone Wash was designated part of the Old Trails Highway and passed through the camp of Oatman, which had grown to the point of having a weekly newspaper, *The Oatman News,* publishing its first issue on December 17, 1914.

In 1915 there were more gold discoveries along the Tom Reed Vein on the Big Jim property to the southeast. When word got out, the gold rush was on and several thousand more people, including promoters, miners and businessmen, arrived in the area.

Life in Early Oatman

The lower-income families lived in tent houses, sometimes connected into multi-tent units, sometimes with roofs reinforced with corrugated tin or other materials. They had no indoor plumbing.

Water was carried in five-gallon cans on a burro, two cans on either side of a pack saddle, from a mine source often as far as one mile away. Drinking water was kept in a hanging olla (pronounced oya, a Spanish word meaning wide-mouthed pot), which was covered with burlap and kept damp.

Food was kept in a homemade cooler covered with burlap. It stayed moist by water dripping from burlap strips attached to a large pan of water on top of the cooler.

Bathing was done in a galvanized wash tub near the kitchen stove. Mesquite wood was gathered for the stove and stored in a fuel box.

Children went barefoot most of the time. Most clothing was homemade, exceptions were overalls, coats and shoes. The children went to a one-room school, sometimes as far away as three miles. They either walked or rode burros.

In the better part of town, the tent houses were somewhat nicer with throw rugs and linoleum on the floors and colored burlap on the walls. There might even be a few fruit trees in the yard.

Before 1916 there was one general store in Oatman and all purchases were charged and paid monthly. Orders were solicited door-to-door early in the morning and delivered later that day.

On paydays the women, usually with a child by the hand, met their husbands when they got off shift and made sure they got home with their paychecks, which the women would cash at the general store. Many single men would cash their paychecks at one of the saloons.

Holidays were family affairs and included Ice Cream Socials held at the school house.

There were no churches in early Oatman but, with the exception of a few drinking problems, there was also virtually no crime.

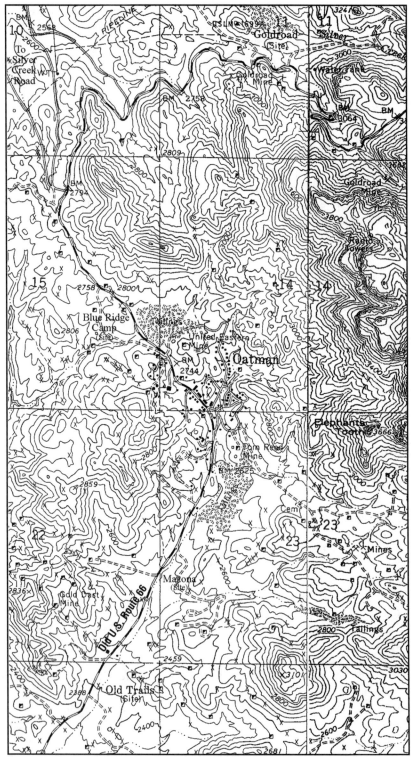

Oatman area map.   Base Maps: U.S.G.S. 7-1/2″ Oatman and Mount Nutt Sheets, 1967.

Oatman — 1916 to 1942

By 1916 four camps, Oatman, Mazona, Old Trails and Gold Road, all along the Old Trails Highway, predominated as centers of activity, the other camps either already gone or to be gone shortly.

Because of its location between two of the largest producing mines in the district, the United Eastern and the Tom Reed, Oatman would quickly become a town.

In February 1916 an Oatman townsite application was filed with an affidavit stating "that there are present within the limits of said township approximately 300 actual residents and not less than 75 buildings, occupied as dwellings by the residents of the townsite. ..."

When the U.S. General Land Office examined the townsite application, it was found that in the Oatman Mining District there were the following townsites in the process of organization located on patented and unpatented mining claims: Tent City, Rice Fraction, Oatman, Oatman City, Mazona, Ryan Addition, Fairview Tract, Bennet, Old Trails, South Oatman, "Camp 49," Times, North Oatman, and Gold Road.

The application for the Oatman townsite was denied because of all of the confusion, and also because the townsite was located on the patented mining claims of the Tom Reed Gold Mining Company.

The 4th of July has always been a big celebration in Oatman. In 1916 there was an auto race from Oatman to Kingman and back, games, fireworks and dancing.

Young mining engineer in a tent cabin in the Oatman area, circa 1916.
Courtesy of the Mohave Museum of History and Arts, Kingman, Arizona.

In the 1916 school year the Oatman School Board had three teachers. In 1917 there were a total of 28 children in the upper class. Most teachers only stayed for one school year.

Continuing about a mile southward from Oatman down Boundary Cone Wash along the Old Trails Highway were the camps of "Camp 49," Mazona, the Carter Subdivision, and Old Trails. These camps, as well as the camps of Times and Gold Road also grew rapidly.

"Camp 49" was the red-light district and was a self-contained community.

Mazona probably formed as an overflow from Oatman and had a large furniture store, general store, saloons, restaurant, lodging houses and a service station that had 50-gallon drums of gasoline for sale to early travelers. In 1918 a gasoline-ignited fire destroyed the business section of Mazona, which was not rebuilt.

The Carter Subdivision had a number of dwellings.

Old Trails, founded in 1915, had more business establishments than Mazona. These included an assay office, a one-room school, an indoor swimming pool, a bottling works that manufactured soda pop, a bakery and a hospital. It was located along the Old Trails Highway and wagon roads which led to mines, both northwest and southeast. In 1925 its post office closed.

In 1916 Oatman's main street was the Old Trails Highway. It was lined with wooden stores, shops, saloons and cafes fronted by wooden sidewalks. Side streets and paths went off the main street, leading to the homes and tent houses of the miners and millmen.

There was a newspaper, a two-room school house with two teachers, churches, library, drugstore, general store and hotels, including the St. Francis with 60 rooms which was built at a cost of $40,000.

Water for home use was delivered in a horse-drawn wagon and later a Model "T" truck, costing 10 cents for 10 gallons. The water was obtained from the de-watering of some of the mine shafts, including the Tom Reed.

By 1917 5,000 to 7,000 people resided in the district, camping at any available place. Each new mining company promoted its own townsite, selling lots on the open market and tent cities thrived.

In July 1917 *The Oatman News* headline was: "War Against Burros," and the article stated in part:

> The citizens of Oatman, especially those who are compelled to work on night shifts and sleep in the mornings, have declared war on the pestilent burros, who seem[s] to infest the camp in greater numbers than ever. These beasts have no owners, no homes, no regular hours.
>
> They destroy a great deal of property, knock over garbage cans in search of food, and constitute a menace to the health and prevent sleep.

Burros were originally brought into the area by prospectors in the 1860s and 1870s and, for various reasons, were released to run free.

In November 1918 the influenza epidemic that swept the nation reached the Oatman area. Of the 380 cases, 26 died in 21 days.

In 1919 Oatman had five excellent hotels, six good eating places, two mercantile stores, two drygoods stores, a drug store and two banks.

On Monday, June 27, 1921 fire completely destroyed one-half of the business district and some 400 residences in Oatman with over $300,000 in damages. The town was quickly rebuilt and soon boasted four modern hotels including the Oatman Hotel, two banks, two large department stores, a picture theater, good restaurants, daily mail service, an ice plant, electric light service, two hospitals, two doctors, three churches and a fine water supply. There were three school buildings. The early 1920s were good times in Oatman and vicinity.

In 1924 there were 10,000 residents in the area, but within two years many of the mines closed or became lessee operations. The population of Oatman and Gold Road dwindled to a few hundred.

In 1926 the Old Trails Highway was designated U.S. Highway 66 and by 1938 it was finally fully paved. During these times of the Model T Ford, Old Trails Highway/Route 66 over Sitgreaves Summit near Gold Road was one of the most feared grades on the entire road because of its twisting, rocky nature. Many travelers spent the night in Oatman so they could drive the hot desert in the cool of the early morning.

When the price of gold increased from $20 to $35 per ounce in 1933, the Tom Reed and Gold Road mines re-opened, and there was an influx of people to the area.

Another large fire on March 20, 1936 burned nearly an entire block to the ground on the west side of Main Street, including the telephone office, barber shop and cafe.

Main street of Oatman, circa 1920s.
Courtesy of the Mohave Museum of History and Arts, Kingman, Arizona.

Children and white burro in the Oatman area, circa late 1930s. The headframe in the background may belong to the United Eastern Mine.
Courtesy of the Mohave Museum of History and Arts, Kingman, Arizona.

On March 18, 1939 Clark Gable and Carol Lombard honeymooned at the Oatman Hotel.

Oatman After 1942

The gold mines of Oatman were closed at the start of World War II as non-essential to the war effort by Executive Order No. L-208. Most of the people left Oatman looking for other employment. The mines never reopened; however, Oatman was still a stopping place along Route 66.

In 1951 Route 66 was relocated, bypassing Oatman, and the town became little more than a ghost town with about 100 residents.

During the 1950s and 1960s Oatman was the setting for several movies including "Foxfire" (1955), "Edge of Eternity" (1959), and "How the West Was Won" (1963).

In 1970 there were nine businesses open in Oatman, including three bars and several antique shops. In 1975 an average of 1,500 tourists visited Oatman each winter weekend. There were gift shops, restaurants, and other tourist-oriented businesses. The population of Oatman was estimated at about 250 mostly retired residents in the winter, and 150 in the summer months.

Over the past hundred years most of the mining properties in the district were sold and bought several times over, each time with the goal of discovering and mining new ore bodies.

Today Oatman is a tourist attraction: a living ghost town with art galleries, antique shops, restaurants, gunfight re-enactments, and the main attraction — the wild burros still roaming the streets looking for a handout.

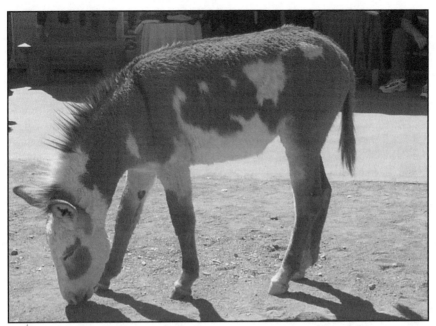

Burros still roam the streets of Oatman. Photo by Ann Ettinger.

Topock (El. 528′)

Topok was a stop on the Atlantic & Pacific Railroad on the east side of the Colorado River, originally called Red Rock. There were other Red Rock stations in southern Arizona, and the stop was re-named Mellen in 1891 in honor of Captain Jack Mellen, a steamer captain on the Colorado River. When Mellen became confused with Needles by telegraphers, the name was changed to Topock in 1916 because it was distinctive. Topock is a contraction of the Mojave Indian word "a-ha-to-pak," meaning water bridge.

Union Pass (El. 3,925′)

Union Pass is located on top of the Black Mountains along Arizona State Highway 68 between Kingman and Bullhead City. Originally the pass lay along an east-west Native American trade route which was in use for several thousands of years.

Captain Lorenzo Sitgreaves was the first American military man to lead an expedition across northwest Arizona to the Colorado River, coming through the pass November 5, 1851.

Between 1864 and the arrival of the A&P in 1883, the Prescott and Mojave Toll Road came through Union Pass, linking Camp Beale Springs with Hardyville along the Colorado River and the ferry (see map, page 56). During this time period there was heavy horse, wagon and stage line traffic carrying men and supplies in both directions. The Union Pass Station offered services to the travelers.

The origin of the name "Union" is not known, but may somehow be connected to the Civil War.

SOUTH OF THE GRAND CANYON AND THE COLORADO RIVER

Ash Fork (El. 5,144')

Named Ash Fork because small ash trees were found where three forks of Ash Creek joined, the town is located 25 miles west of Williams along I-40. Before the Atlantic & Pacific Railroad arrived in October 1882, the original settlement was built as a freight and stageline depot at the junction of the Beale Wagon Road and the wagon road south to Prescott.

When the A&P arrived a siding was constructed, and by November 1882 a station and depot were in use for passengers and freight destined for as far south as Phoenix.

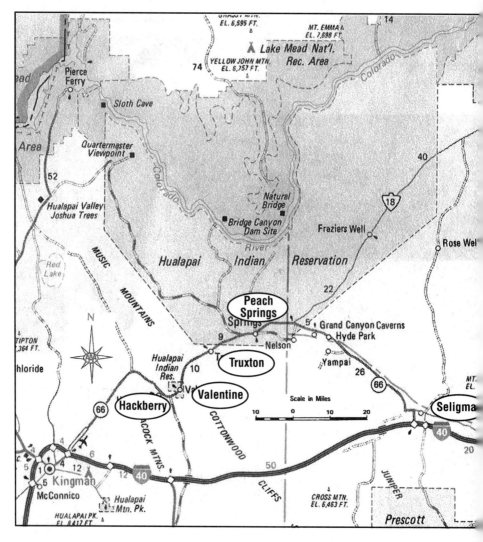

Places on the map — South of the Grand Canyon.

On April 25, 1885 fire destroyed nine buildings, more than half the town's businesses, including the hotel and most of the saloons along "Whiskey Row." That summer most of Ash Fork's men left for the mining rush to Cataract Canyon.

Another fire in January 1886 destroyed two hotels, a bakery and the meat market. The town burned again in 1893 and a new town was built on the other side of the railroad tracks, including a depot and eating house.

The name of Ash Fork was changed to Ashfork in 1894, and then back to Ash Fork in 1950.

In 1905 a new depot was built along with a new hotel and Harvey House a half mile west of the existing station. Opening in the fall of 1906, the Escalante Hotel had 30 sleeping rooms and its own electricity generating plant. The hotel remained open until 1948.

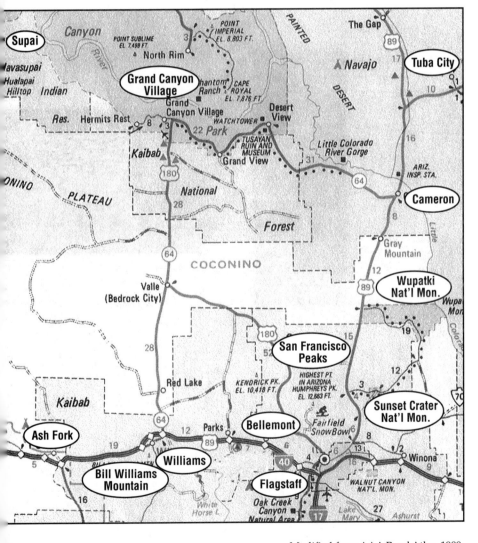

Modified from AAA Road Atlas, 1989.

In 1960 the Santa Fe realigned its tracks, and Ash Fork suffered a major economic setback compounded when U.S. Route 66 was replaced by I-40.

Today Ash Fork, with a population of 530, lies one mile north of the junction of I-40 and U.S. 89, which continues south to Prescott and Phoenix and is known as the "Flagstone Capital of the United States."

Bellemont (El. 7,132')

Located about 10 miles west of Flagstaff, Bellemont was originally called Volunteer because of its proximity to 8,047 foot Volunteer Mountain, three miles to the southwest. Before the Atlantic & Pacific Railroad arrived, the site was a stage and relay station on the northern Arizona route.

In 1876 Walter Hill homesteaded the Volunteer Springs Ranch, which became one of the largest sheep ranches in the Territory.

In 1882 when the A&P arrived, the town's name was changed to Bellemont to honor Miss Belle Smith, daughter of F.W. Smith, general superintendent in charge of railroad construction in the early 1880s.

There were sheep and cattle ranches near Bellemont. The Hill Ranch was located a mile south of Bellemont, and the A&P transported thousands of sheep to market over the years.

The Arizona Lumber Company built a branch sawmill near Bellemont.

In 1898 the Santa Fe built a tie-treating plant in Bellemont which burned in December 1902 and again in July 1906. The plant was not rebuilt following the 1906 fire. During this same time, about two miles west of Bellemont, the Santa Fe operated a cinder pit for railbed ballast, employing up to 50 men.

At the beginning of World War II, the U.S. Government established the Navajo Ordnance Depot which covered thousands of acres of the early Hill Ranch to the south. In April 1942 construction began on 800 building units, 50 miles of railroad track, and 185 miles of paved roads. The construction payroll peaked at 8,000 men, many of whom commuted from Flagstaff by bus. The Victory Bus Line operated seven daily round-trips from Flagstaff.

In 1962 the name was changed to Navajo Army Depot and was operated by the Arizona National Guard. In October 1993 ownership was transferred to the Arizona National Guard, the name changed to Camp Navajo, and it was used as a training base and for storage of Air Force materials.

Today Bellemont is a small community along I-40.

Bill Williams Mountain (El. 9,256')

William Sherley Williams was born in North Carolina on January 3, 1787. He served as an itinerant preacher for nine years, spent 12 years on the frontier and then seven years as a plainsman and mountain man. He was tall, gaunt, redheaded and fairly well-educated. Williams was a hunter, trapper and guide in the Southwest, but by 1837 he was in northwest Arizona living off the land and trapping beaver. In late 1848 Williams was hired as a guide for the fourth Fremont expedition to California. The expedition got caught in winter storms in southern Colorado and went no further. Williams and Dr. Ben Kern, a traveling companion, were killed by a war party of Ute Indians on March 14, 1849. Bill Williams was 62 years old.

In 1851 Richard Kern, Dr. Ben Kern's brother, named the double-peaked volcano near present-day Williams "Bill Williams Mountain."

Today the mountain has downhill and cross-country ski facilities, and also offers camping, hiking and fishing recreation.

Cameron (El. 4,200')

Cameron, the eastern gateway to the Grand Canyon, is located 51 miles north of Flagstaff on U.S. 89. The town was named for U.S. Senator Ralph Cameron, last territorial delegate from Arizona to the U.S. Congress.

In 1911 a bridge was constructed across the Little Colorado River, a store was built and settlement began.

A trading post was built on the south rim of the river by Hubert and C.D. Richardson in 1916. Over the years a motel and restaurant were added and the trading post expanded.

Today the trading post is operated by C.D. Richardson's grandnephew, and has one of the largest inventories of Native American craft items and souvenirs in northern Arizona. Cameron has a population of approximately 1,000.

The original Cameron Trading Post, built in 1916, is now used as a gallery. Photo by L.J. Ettinger, March 2005.

Cataract Canyon and Havasu Creek and Canyon (El. 1,500')

Cataract Canyon is located 20 miles west of Grand Canyon Village. Cataract Canyon becomes Havasu Canyon south of Supai and Havasu Creek flows northward passing close to Supai, and then into the Colorado River.

There are three beautiful waterfalls (Navajo, Havasu, and Mooney) along Havasu Creek north of Supai.

FLAGSTAFF (El. 6,905')

Legend tells that in 1871 Edward Whipple operated a saloon near Antelope Spring close to the well-traveled Beale Wagon Road, which had been surveyed in 1859. In 1876 Thomas F. McMillen, the second settler to arrive, brought a herd of sheep and made camp north of present-day Flagstaff. A few months later an advance party of scouts for a wagon train of settlers camped near the spring and found an open valley with a lone pine tree. Stories vary, but sometime in mid-1876 the tree was stripped and an American flag was raised.

The original settlers numbered less than 20 in 1880.

The spring, originally called Antelope and then Flagstaff, was changed to Old Town Spring with the coming of the Atlantic & Pacific Railroad.

In October 1881 because of the A&P construction, Old Town Spring was one of the liveliest towns in Arizona. There were railroad graders, gamblers, and "fast people" who followed the boom times. Downtown Old Town Spring grew. There was a post office and the population of about 200

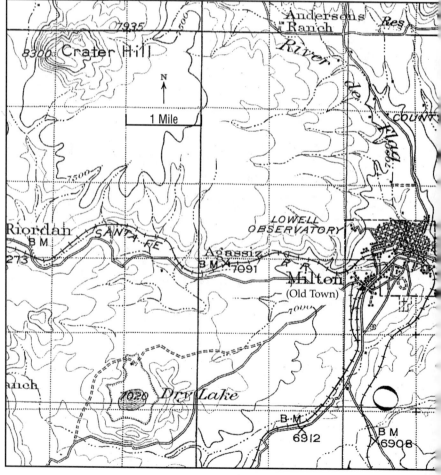

U.S.G.S. Flagstaff, AZ. 30' Quad Sheet, 1912.

were housed in 20 frame buildings and some 20 tents. The population doubled on Saturday nights, and the town ran full blast from Saturday night to Monday morning. Fights on the street could involve 20 or 30 men with gunfire in all directions.

Two months later Old Town (later called Milton on maps) was all but deserted, the railroad construction crews having moved westward. In 1882 there were 10 buildings at Old Town.

In the summer of 1882 trains began to arrive at a new depot named Flagstaff, which consisted of two box cars on a side track about a mile northeast of Old Town. New businesses opened near the depot which became the area's center of activity.

The area had an abundance of nearby Ponderosa pine and Flagstaff became the center of the lumber industry. The Arizona Lumber & Timber Company was the largest operation, and they located a sawmill west of town where Old Town had been before it burned in 1884. Stockyards were built and Flagstaff became a shipping point for cattle and sheep. The Flagstaff Hotel opened in March of this same year.

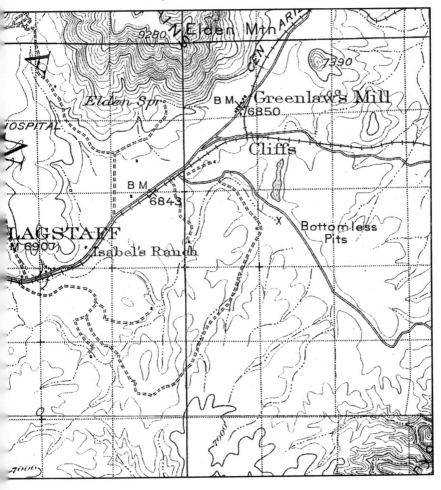

By 1886 Flagstaff was the biggest city on the main link between Albuquerque, New Mexico and the Pacific Coast. In February a fire in a Chinese restaurant spread and destroyed 30 buildings.

In 1887 both Flagstaff and Old Town enjoyed a building boom, but a fire destroyed the Arizona Lumber Company's sawmill. In October of that year work began on a passenger station which was completed in mid-1888. The stockyard was relocated two miles east of town, which eased the problem of loose cattle roaming Flagstaff streets, or the danger of a stampede when loading cattle cars.

In July 1888 another fire destroyed 18 structures in Flagstaff's business district.

The Arizona Sandstone Company began operations about a mile east of the depot in 1888, and by 1890 was shipping four railroad cars of sandstone west to the Pacific Coast and east to Kansas City.

There were two fires in July 1889, one of which destroyed the A&P depot. The depot was rebuilt as a stone structure.

The population of Flagstaff in 1890 was 963, many of whom worked in the sawmill west of town, the sandstone quarry or for the railroad.

In 1891 Flagstaff had grown to 1,500 and had become the seat of newly-created Coconino County, and by 1892 the original flag pole was gone.

In November 1892 two fires destroyed the Opera House and the "wooden row" on Front Street.

The local residents enjoyed picnics during the summer and ice skating and sleigh riding in the winter months. Saloons charged 12-1/2 cents for a drink. The depot was the community social center during the week. When

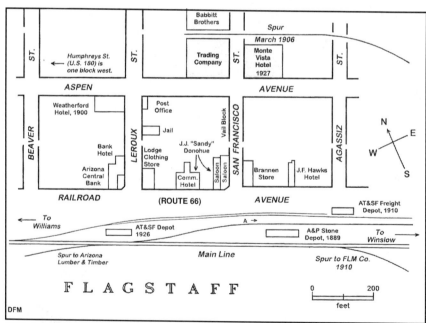

1901 Survey map, with some later additions. "Railroad Avenue" was renamed "Santa Fe Avenue" in 1927, and later decreed "Route 66."   Modified from Myric, 1998.

178

the daily California Limited arrived from either direction, the residents came from all directions to see the activity. On Sundays, cowboys held roping contests on vacant lots.

In 1894, because of its pure air, Dr. Percival Lowell chose Flagstaff as the site to build his now-famous Lowell Observatory. A 40-inch lens arrived in 1909, at the time the third largest in the United States. Mars was studied and the planet Pluto was discovered in 1930. Later studies included the expanding universe.

In 1898 fire again destroyed the Arizona Lumber Company's sawmill.

The Northern Arizona Normal School, a teacher's college, opened in 1899 with 23 students. It became Northern Arizona University in 1966.

Flagstaff was incorporated as a city in 1928.

World War II brought an influx of about 8,000 construction workers and their families to Flagstaff to construct the Navajo Ordnance Depot, about 12 miles to the west near Bellemont.

Flagstaff is center to many scenic and recreational areas including Grand Canyon, Sunset Crater Volcano National Monument, Wupatki National Monument, Oak Creek Canyon, Walnut Canyon National Monument, San Francisco Peaks, Montezuma's Castle National Monument and Meteor Crater.

Today, Flagstaff has a population of nearly 53,000 and continues to grow.

Grand Canyon Village (El. 6,900')

In the mid-1880s several small camps were established along the south rim. Tourists were picked up at the Atlantic & Pacific Railroad station in either Flagstaff or Williams and traveled to the camps by horse or stagecoach for their stay at the Grand Canyon.

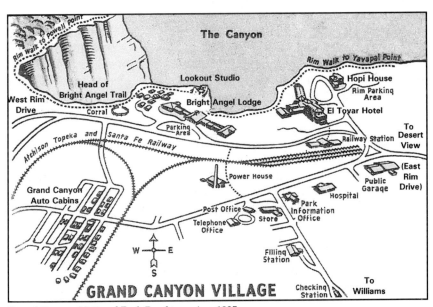

Grand Canyon National Park Brochure, circa 1935.

In 1891 Ralph Cameron opened the Bright Angel Trail for tourists, but the toll fees were so high that the visitors went to other camps.

Realizing the tourist business potential, in 1901 the Santa Fe Railroad built a 65-mile spur line from Williams to an area along the South Rim that would become Grand Canyon Village. Along with the Santa Fe came their long-time business associates, the Fred Harvey Company, to develop tourist facilities along the South Rim close to the station.

Grand Canyon Village had its beginning in 1905 when the Fred Harvey Company opened several lodges including the El Tovar Hotel and gift shops.

The Santa Fe and Fred Harvey companies did not want competition of any kind, and there were continual hostilities and feuding with other businessmen, including the Kolb Brothers with their photography business, and Ralph Cameron, who controlled the Bright Angel Trail. Litigation with Cameron lasted for years, finally forcing Cameron out of the area.

When Grand Canyon National Park was created in 1919, park headquarters were established at Grand Canyon Village which became the center of activities on the South Rim.

Over the years improvements have been made and additional facilities have been built by the National Park Service and the Fred Harvey Company to accommodate the large number of tourists. Today more than four million people visit the Grand Canyon each year.

Hackberry (El. 4,000')

Hackberry is located 23 miles west of Peach Springs. Originally called Gardiner Spring by Lt. Edward Beale in 1857, the name was changed to Hackberry Spring in 1874 when four prospectors found a rich vein of silver ore about one-half mile west of the spring. The name was derived from hackberry trees found beside the spring. In the summer the hackberry trees attract hoards of birds that feed on its berries.

Mining operations began in 1878 and a camp was built near the spring.

When the Atlantic & Pacific Railroad arrived in 1882, the town of New Hackberry was established about 1.5 miles north of the mining camp near the railroad station, quickly grew to a population of over 400, and became a shipping center to supply the mining camps in the area.

In 1897 there were saloons, three hotels, two stores, one bath tub, and an Indian school.

Hackberry became one of the important cattle shipping points along the Santa Fe. Thousands of cattle were shipped annually to Bakersfield, California for fattening or to the stockyards in Kansas City. During the shipping season cowboys filled the hotels, saloons and cafes.

In April 1898 a fire destroyed the railroad station.

Strawberries grown in Truxton Canyon were shipped by rail every spring.

In 1906 the Santa Fe opened a gravel pit near Hackberry which was used for roadbed ballast along the line.

Hackberry was a stop for trains for fuel. When steam locomotives were replaced by diesel locomotives in the 1940s, the Santa Fe abandoned Hackberry.

U.S. Route 66 passed about a mile north of Hackberry on the northeast side of the railroad tracks.

Peach Springs (El. 4,788')

Peach Springs is located along the Santa Fe main line and U.S. Route 66, 120 miles west of Flagstaff and 50 miles east of Kingman.

Three groups of springs found in a canyon were known to the Hualapai for hundreds of years. They were originally named "St. Basil's Wells" by Fr. Garcés on June 15, 1775.

Peter Skene Ogden of the Hudson's Bay Company explored northwest Arizona in 1829 and camped at the springs, naming them "Young Springs" after Edwin Young, a member of the party.

On September 17, 1858 Lt. Edward Beale camped at the springs and called them Indian Spring. On April 23, 1859 the Udall party camped at the springs, calling them Hamphill Camping Ground Springs.

In the early 1860s Mormon missionaries stopped at the springs and a child planted some peach pits, which in time, grew into trees.

The Atlantic & Pacific Railroad tracks reached the area of the springs in January of 1883. The station was called "Peach Springs" and designated as their fourth division point. The railroad constructed a yard, a roundhouse, shops, a coal chute, and a 50,000 gallon water tank three miles southwest of Peach Springs. A settlement was laid out in the canyon for the work crews.

Also in January 1883 the Hualapai Indian Reservation was created covering the area of Peach Springs.

The A&P needed the water from the springs and there was controversy over control of the springs for years, followed by an extended lawsuit by the Hualapai in the late 1920s.

Fred Harvey Company owned a nearby ranch which supplied beef and dairy products to their eating houses.

In the late 1880s Peach Springs was an early point of departure for the Grand Canyon stages. Tourists stayed in the rooms at the eating house, which was destroyed by fire in 1891.

In the mid-1890s the Grand Canyon Lime and Cement Company opened their Nelson Mine, a limestone quarry five miles east of Peach Springs, which has seen a large production over the years.

In 1938 the Peach Springs settlement was designated as the tribal capital of the Hualapai.

Today Peach Springs, with a population of about 600, is the trading center and headquarters for the Hualapai Indian Reservation, which covers nearly a million acres between the town and the Colorado River. Peach Springs is also the starting point for access to the western part of the Grand Canyon.

San Francisco Peaks (El. 6,000' to 12,633')

Located north of Flagstaff, the San Francisco Peaks are the center of extinct volcanic activity dating about 2.8 million to about 200,000 years ago, covering an area of about 3,000 square miles.

San Francisco Peaks, looking north.   Photo by Ann Ettinger.

The "San Francisco Peaks" were likely named in 1629 by Franciscans at Orailie to honor the founder of their order, St. Francis of Assisi.

The three highest peaks in Arizona are Mt. Humphreys (El. 12,633'), Mt. Agassiz (El. 12,356') and Mt. Fremont (El. 11,969'). These peaks are the eroded remnants of a stratavolcano, similar to Mt. Fuji, Mt. Shasta, Mt. Lassen, or Mt. St. Helens.

The three peaks have been a landmark, visible for miles, and important in the folklore of both the Navajo and Hopi Indian tribes.

Seligman (El. 5,242')

Seligman is located along Route U.S. 66, 25 miles west of Ash Fork.

It was originally called Prescott Junction when a settlement was built where the new railroad spur line from Prescott joined the Atlantic & Pacific Railroad in 1886. A post office was established November 9, 1886 and a Wells Fargo Station in 1887. The spur line was abandoned and the rails removed after a few years in favor of a branch junction at Ash Fork.

In time the small community came to be known as Seligman after the Seligman brothers, New York bankers, who were connected with the A&P. They also owned the local Hash Knife Cattle Company.

When the A&P reorganized in May 1897, Williams was abandoned as its western terminus in favor of Seligman. In July 1897 the Santa Fe Railroad emerged from the reorganization, and the roundhouse at Williams was dismantled and moved to Seligman.

Fred Harvey Company built the Havasu Hotel about 1905, and the dining room closed about 1930.

Today Seligman has a population of about 450.

Supai (El. 3,195')

Supai is located 65 miles northwest of Peach Springs in Havasu Canyon below the southern rim of the Western Grand Canyon.

The village of Supai serves as the governmental center of the Havasupai Indian Reservation.

Automobiles must be parked at Hualapai Hilltop. Supai can only be reached by an eight-mile journey to the canyon floor by horseback, helicopter, or on foot down a precipitous trail. Today's population is about 400.

Sunset Crater National Monument (El. 8,000')

Located 12 miles north of Flagstaff on U.S. 89, Sunset Crater Volcano is the youngest volcano in Arizona, erupting about A.D. 1064 and was sporadically active for nearly 200 years. Its truncated cone is 1,000 feet high and is surrounded by fields of cinders, lava flows and splatter cones. Lava tubes underlie the surface, some of which are formed ice caves.

In the 1880s ice from these caves was used in homes, saloons and other businesses in Flagstaff.

The crater was named by Major John Wesley Powell in 1892 because of the cinder colors which graded downward from the summit ranging from shades of yellow, orange, red, deep red to black on the lower part of the cone.

The Sunset Crater National Monument was established in 1930.

Truxton (El. 3,880')

Truxton is located nine miles southwest of Peach Springs along the Santa Fe main line. Springs were long known and used by the Yavapai in Truxton Canyon, which forms the boundary between the Colorado Plateau on the east and the Basin and Range Province on the west.

The town was named Truxton by Lt. Edward Fitzgerald Beale on his survey in 1858 in honor of his wife, whose maiden name was Truxton.

In 1883 the Atlantic & Pacific Railroad put in a siding, a large pump and a water tank at Truxton Springs which, for many years, served as an important watering place on the railroad.

The opening of U.S. Route 66 gave Truxton an economic boost.

Tuba City (El. 4,936')

With over 30 springs in this area it was a natural stopping place for the Hopi, Havasupai, Navajo and Paiute for hundreds of years. Some of these early travelers remained and cultivated crops.

The area was visited by Fr. Francisco Garcés in 1776, and Mormon explorer Jacob Hamblin passed through in the 1850s and 1860s. The area was settled in 1875 and a townsite was laid out in 1878 using blocks of stone from nearby prehistoric sites.

Originally called Tuba after Hopi Chief Tuve by the Mormons, a post office was first opened in 1884. The town grew and in 1894 assumed its present name. In 1903 it was discovered that the town was located on Indian land, and the U.S. Government bought out the improvements and the Mormons moved away. Tuba City later became the headquarters of the Western Navajo Agency.

Today Tuba City has a population of 8,225.

Valentine (El. 3,800')

Valentine, an unincorporated village, is located four miles east of Hackberry along U.S. Route 66. It was originally called Truxton Canyon Subagency and was established in Truxton Canyon in 1898.

In 1900 an Indian school was built on 666 acres of land which had been set aside. In 1938 the Peach Springs Hualapai settlement was designated as the Tribal Capital. When the Indian school closed, the Truxton Canyon Post Office was moved a mile away, off of Indian land, and renamed Valentine after Robert G. Valentine, Commissioner of Indian Affairs from 1908 to 1910.

Williams (El. 6,752')

The first men in the area were Sam Ball and John Vinton who arrived in 1876. In 1877 their interests were bought out by cattleman Charles Thomas Rogers who homesteaded a 160-acre ranch in the area that is now Williams.

In 1881 the Atlantic & Pacific Railroad reached the area of Rogers ranch and Louis Kingman, superintendent in charge of building the railroad, set up his headquarters on the ranch along with a large railroad camp and supply yard of construction material.

The A&P would challenge Roger's ownership of his ranch and, after ten years of litigation, prevailed except for several lots in Williams.

The settlement, near the base of Bill Williams Mountain, was called Williams, became a railroad division point until 1885, and an eating stop on the A&P. With the loss of the division terminal, most of the tent houses and shanties were taken down as the railroad employees moved elsewhere.

Early Williams' economy was from the railroad, lumber and livestock, and the town grew into a typical small western town with many saloons, sawmills, and cattle yards.

Williams had more fires than most Arizona communities. In July 1884 two fires in one week occurred in Williams, the second one destroying the business district.

In 1887 Williams again became a division point on the A&P and a new roundhouse was constructed.

In March 1889 another fire in a saloon swept across the business section destroying three saloons, one restaurant, and an emporium.

The Saginaw Sawmill began operation in 1893 southwest of town.

Another fire swept through Williams in September 1897, and a month later the A&P division was moved to Seligman.

The most destructive fire in Williams occurred in July 1901. With a stiff breeze and lack of water, the fire destroyed most of the business section including the bank, opera house and newspaper office.

In 1901 the Santa Fe Railroad constructed a spur line from Williams to the South Rim to provide tourist access to the Grand Canyon.

Tourist facilities were built by the Santa Fe and Fred Harvey Company at the end of the line and named Grand Canyon Village. Tourists quickly provided most of the traffic on the line and in 1907-08 the Santa Fe improved their facilities at Williams, with several miles of new track in their yard, a new 85-foot turntable, and a six-stall roundhouse. Fred Harvey built a station-hotel-restaurant, The Fray Marcos Hotel, which opened on March 10, 1908.

Fire struck again in the fall of 1908 and again in February 1911.

Williams became known as the "Gateway to the Grand Canyon." Passenger service operated until 1968.

In 1989 the Grand Canyon Railway from Williams to the South Rim was reinstated, and in 1992 carried about 100,000 passengers.

Today Williams is a resort town with a population of about 2,500.

Wupatki National Monument (El. 4,900')

Wupatki National Monument is located 30 miles north of Flagstaff on U.S. 89. After the eruption of Sunset Crater, 20 miles to the south, in about A.D. 1064, there was increased rainfall in the area and farming became productive. The original inhabitants were Anasazi, who remained in the area until about A.D. 1300. Artifacts found include turquoise and shell jewelry.

Lt. Lorenzo D. Sitgreaves, on his expedition through northwest Arizona, camped at Wupatki in 1851.

Today one can see dozens of ruins, dated at A.D. 1100 to 1225, within the 55 square-mile monument. The largest ruin is the "Wupatki" (Hopi for "big house") pueblo, home to 300 people with the "Long-Cut House," containing more than 100 rooms in addition to a nearby amphitheater, ball court and "blow hole." Other ruins include "The Citadel," "Nalakihu," "Lomaki" and "Wukoki."

The area was set aside as a national monument on December 9, 1924.

The ruins are still visible today. The Lomaki and nearby stone structures are built on limestone outcrops along narrow creek beds. The Anasazi grew corn where the creek bed widened, and built small dams to impound water from rain and melting snow.

Wukoki ruins, Wupatki National Monument.   Photo by Ann Ettinger.

NORTH OF THE GRAND CANYON AND THE COLORADO RIVER — "THE ARIZONA STRIP"

Colorado City, Arizona and Hildale, Utah (El. 5,000')

Originally known as Short Creek, the area was settled in 1913 by a cattle rancher in northwest Arizona just south of the Arizona-Utah state line.

Polygamy was disavowed and banned by the Church of Jesus Christ of Latter-Day Saints (LDS) in 1890. Those Mormons who continued to practice polygamy went into hiding, many relocating to the "Arizona Strip," a relatively isolated area north of the Grand Canyon and south of the Arizona-Utah state line.

Short Creek became a stronghold for the Lee's Ferry polygamists who were excommunicated from the LDS Church in 1935 after refusing to sign an oath against polygamy. Within a few years thereafter, several polygamists formed the Fundamentalist Church of Jesus Christ of Latter-Day Saints. The Fundamentalists located in Short Creek because of its isolation and proximity to the state line, which could easily be crossed if there were trouble.

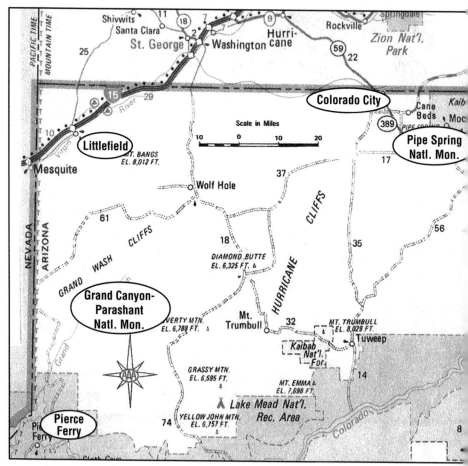

Places on the map — North of the Grand Canyon.

Over the years Short Creek was periodically raided by local and federal law enforcement officers, resulting in jail and prison sentences for some of the polygamist men. Sometime after a 1953 massive police raid, the Fundamentalist Church renamed Short Creek, Colorado City, Arizona and Hildale, Utah in order to avoid any connection with the "Short Creek Raid."

In recent years there have been power struggles within the Fundamentalist Church, litigation and criminal prosecutions.

Fredonia (El. 4,750')

Fredonia was founded in 1865 by a group of Mormon settlers fleeing Utah in search of religious freedom from the federal laws prohibiting polygamy. The name Fredonia comes from the combination of the English word "free" and "donia" (the Spanish word for woman), meaning "free woman."

Fredonia was incorporated in 1956.

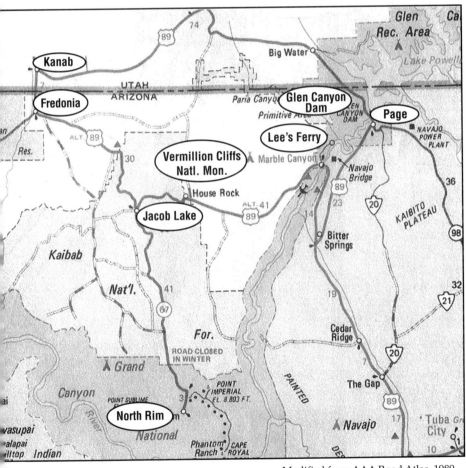

Modified from AAA Road Atlas, 1989.

Glen Canyon Dam (El. 3,700')

Bureau of Reclamation engineers and geologists selected the future Glen Canyon dam site between 1946 and 1948. Site criteria included a large upstream basin, canyon walls and bedrock that were strong and stable, and a nearby source of good rock and sand.

In 1956 Congress passed the Colorado River Storage Act which provided for construction of four major water storage and hydroelectric power generating dams on the upper Colorado River and its tributaries. Permitting was easier then than it is today because it was prior to the law requiring environmental impact statements for major construction projects.

The location chosen was inaccessible, and construction crews had to drive some 200 miles to cross from one side of Glen Canyon to the other.

Under direction of the U.S. Bureau of Reclamation, the first foundation blast occurred October 15, 1956 on the Glen Canyon Dam Project.

New highways were built from Kanab, Utah and in Arizona from Bitter Springs north to the Page townsite and the Colorado River. By 1957 the newly constructed town of Page was occupied, and the Glen Canyon Bridge spanning the Colorado River was in use in 1959.

Over five million cubic yards of concrete were poured day and night from 1960 through 1962. The dam was completed in 1964. Turbines and generators were installed from 1963 to 1966.

The plant generates more than 1.3 million kilowatts of electricity from eight generators which require more than 15 million gallons of water each minute.

Lady Bird Johnson dedicated the dam on September 22, 1966.

Lake Powell, 560 feet deep, took 17 years to reach its full level at 3,700 feet. When full, in its 186 miles in length with an estimated 1,950 miles of shoreline, Glen Canyon including more than 200 side canyons is the second largest man-made reservoir (Lake Mead is the largest). Today Lake Powell, most of which lies in southern Utah, is within the Glen Canyon National Recreation Area established in 1972, and attracts 2.5 million tourists annually.

Glen Canyon Dam created a new Colorado River through the Grand Canyon. Once filled with silt and sediment, the river now drops its load in Lake Powell. The temperature of the water below the dam has changed as have the fish and plant life. Since 1996, with evidence of deteriorating conditions along the benches of the Colorado River, the floodgates of the dam have been opened several times for one to two days to flush the river.

Grand Canyon - Parashant National Monument (El. 2,000' to 7,000')

This National Monument was established in 2000 and covers more than one million acres north of the Grand Canyon and east from the Arizona-Nevada state line. The boundary between the Basin and Range and the Colorado Plateau provinces passes through the monument, showing a range of rock formations from 1.7 billion year old pre-Cambrian granites to 400 million year old limestones.

Archaeological sites, including petroglyphs, pithouses and villages, indicate Indian presence going back 7,000 years. This includes small bands of hunter-gatherers of the Desert Cultures and later Anasazi and Southern

Paiute Cultures. Historic sites include abandoned Mormon homesteads, ranches and mining camps.

Animal life includes bighorn sheep, coyotes, mule deer, wild turkey, the Kaibab squirrel, and the California Condor.

The area has no paved roads, services or developed recreation sites.

Jacob Lake (El. 7,921')

Located on the Kaibab Plateau between the North Rim of the Grand Canyon and Fredonia, Jacob Lake was named after Jacob Hamblin, a Mormon missionary who explored the area, including the Little Colorado River country, for settlement sites in the 1860s and 1870s. He established the Mormon Road for emigrants traveling from Utah into northern Arizona.

Jacob Lake is a small, shallow lake that nearly always contains some water and is used by stockmen. Today there is a small settlement and campground nearby.

Kanab, Utah (El. 4,925')

Kanab, an Indian word meaning "a willow basket used to carry an infant on the back of its mother," was first settled by Mormons in 1858. Because of attacks by Paiutes over the next 12 years, attempts to colonize the area were discouraged.

In 1870 a fort at Kanab was a bustling center of activity, becoming a focal point for local pioneering, exploration, trading, and base of operations for the U.S. Geological Survey.

Beginning in the 1920s, and because of the scenic background, hundreds of movies and television episodes have been filmed in the area.

Kanab has always been a cattle town and received an economic boost in 1956 with construction of the Glen Canyon Dam.

Today Kanab, the gateway to the North Rim of the Grand Canyon, Zion and Bryce National Parks, and the Coral Pink Sand Dunes has a population of 3,300.

Lee's Ferry (El. 3,170')

Lee's Ferry is located 15 miles south of the Utah-Arizona border, 15 miles below the Glen Canyon Dam, and just downriver from where the Paria River joins the Colorado River in Marble Canyon.

Between 1860 and 1863 Brigham Young directed members of the Church of Jesus Christ of Latter-Day Saints to settle along the Little Colorado River and also in other places in northern Arizona.

To enter into Arizona, the Colorado River would have to be crossed. In 1860 Mormon missionary and frontiersman Jacob Hamblin arrived at the mouth of the Paria River while searching for a safe crossing of the Colorado River, but was unsuccessful in his first attempt.

In March 1864, Hamblin returned to the junction of the Paria and Colorado rivers. He and 15 of his men built a raft and made the first successful crossing at a point originally called "Paria Crossing," which would later become Lee's Ferry. It would be learned later that this was the only place between Moab, Utah and Pierce Ferry, Arizona where a wagon could be driven to both sides of the Colorado River. In 1865 a small Mormon settlement was established at the mouth of the Paria River.

Over the next few years the Paiutes raided the Mormon settlements and by 1869 the Mormons posted guards at the crossing. A small stone building and corral were erected and called "Fort Meeks."

Major John Doyle Lee, an adopted son of Brigham Young, was alleged to have led the Mountain Meadows Massacre on September 7, 1857, when wagon trains of the Fancher Party from Arkansas and Missouri were attacked some 30 miles west of Cedar City, Utah by Mormons and Indians. Lee claimed he was not part of the massacre, but he was excommunicated from the Mormon Church.

Later he was asked by the Church to establish and operate a ferry that could be used by Church emigrants traveling south on colonizing missions.

Lee spent much of his time evading federal law enforcement officials and visiting his numerous wives, who were scattered across southern Utah.

In 1871 Lee became the first "permanent" resident of the area just downstream from where the Paria River joins the Colorado River. He established a small ranch called "Lonely Dell" on the river valley floor, and the Lee family raised livestock, farmed and planted an orchard of pear, apricot, peach and plum trees. Lee also built and operated a ferry.

Major John Wesley Powell, on his second expedition down the Colorado River, camped at Lee's Ferry between September 1871 and August 1872.

Wagon roads were constructed over rocky outcrops, later described as "bone-jarring and wagon breaking," on both sides of the river. Ferry operations began in 1873.

Lee, a wanted man, was captured by United States Marshals in 1874, tried twice, found guilty and executed by firing squad at Mountain Meadows in 1877. After Lee's death in 1877, Emma, one of his 19 wives, kept the ferry operating until the Mormon Church bought the operation later the same year.

During the 1870s and 1880s thousands of Mormon emigrants bound for Arizona used Lee's Ferry as a crossing point. Travelers faced muddy banks, a fluctuating, sediment-filled, dangerous river and a ferry boat that had been involved in several accidents.

Ferrage fees for Mormon travelers were $2.00 per wagon, $1.00 per horse and rider, and 25 cents per head of stock, while non-Mormons paid about 50 percent more.

Over the years a number of rock and log buildings were built and used for the farm and ferry operators until 1928.

The largest and most historically significant building found at Lee's Ferry is the Lee's Ferry Fort, built in 1874 for protection during the Navajo uprising, and later used as a Navajo trading post.

Between 1910 and 1913, a little upriver from Lee's Ferry, the American Placer Company, headed by Charles H. Spencer, unsuccessfully attempted to extract gold from the Chinle Shale. Several buildings, a large boiler and remains of a sunken steamboat can still be seen at the location.

In 1909 the Church sold the ferry to a cattle company. In 1916 Coconino County took over the ferry and ran it until Navajo Bridge was opened in 1929.

Nearby, other buildings were constructed in 1921 by the U.S. Geological Survey where Colorado River flow was measured as part of the Colorado River Compact.

Today Lee's Ferry and Lonely Dell Ranch are Historic Districts administered by the National Park Service. Lee's Ferry is the launching point for thousands of tourists who raft through the Grand Canyon each year, and also is a world-class trout fishing area.

Littlefield (El. 1,600')

Located in the extreme northwest corner of Arizona along the Virgin River, Littlefield was settled in 1864 by a group of Mormon settlers led by Henry W. Miller. Originally called Millersburg, it was later renamed Beaver Dams because of the abundance of beaver.

In 1867 the Virgin River flooded and the area was abandoned. In 1877 new settlers arrived, and a permanent settlement was named Littlefield because of the numerous small farms in the area.

North Rim (El. 8,255')

The earliest occupants at the North Rim were the Anasazi, whose structural ruins are found at the Walhalla Overlook near Cape Royal.

Development of the North Rim of the Grand Canyon lacks much of the colorful history as that of the South Rim.

The Union Pacific Railroad obtained concessionaire rights to the North Rim in the mid-1920s. Architect Gilbert Stanley Underwood was retained to design a rustic national park lodge that included a massive Spanish style exterior with a high front, topped by an observation tower which was built in 1928. In 1932 the lodge burned, flames spreading from below the kitchen which engulfed the structure within minutes.

A new lodge was built in the summer of 1937 using the same floor plan as the original lodge. The new structure had sloping roofs which were better able to shed heavy winter snows, but retained Underwood's view of the canyon.

North Rim elevations range from 8,000 to 8,800 feet and can result in harsh winters with much snowfall. Therefore, the road to North Rim facilities is only open from approximately mid-May through October.

Early visitors to the North Rim had a long trip through Jacob Lake and then 44 miles south to the lodge. As an added attraction, the White Motor Company Buses were met by singing college-age employees.

To reach the North Rim from the South Rim today is a 215-mile drive which takes five hours. The route passes through the Navajo Reservation, the Painted Desert and the Kaibab National Forest. Only about 10 percent of the visitors to the Grand Canyon visit the North Rim.

Page (El. 4,280')

Page is a community built in 1957 to provide homes and community life for construction workers on the Glen Canyon Dam. Named after John Chatfield Page, Commissioner of Reclamation from 1937 to 1943, who devoted many years to development of the Upper Colorado River.

Today Page has a population of 6,800 and is a jumping-off point for the Glen Canyon National Recreation area. The Navajo coal-fired powerplant is located a few miles east of Page.

Pierce (Pearce) Ferry — Site (El. 1,200')

In December 1876 Harrison Pearce built a ferry crossing the Colorado River near the western edge of the Grand Canyon for use by Mormons wishing access from southern Utah to the mines along the Colorado River and in northwest Arizona. The ferry settlement disappeared under the waters of Lake Mead after construction of Hoover Dam.

Pipe Spring National Monument (El. 5,000')

The waters of Pipe Spring were used by the Anasazi and Kaibab Paiute for over 1,000 years. They gathered grass seeds, hunted and raised crops near the springs.

In 1858 Mormon pioneers, led by Jacob Hamblin, camped at Pipe Spring while on an exploration trip to find a crossing on the Colorado River and to make peace with the Indians. In 1863 a small cattle ranch, owned by the Mormon Church, was in business. In 1869 Winsor Castle Fort was built over the main spring and a large cattle operation was established. During the 1880s and 1890s the Pipe Spring ranch served as a refuge for polygamist wives.

In 1907 the Kaibab Paiute Indian Reservation was established which surrounded the privately owned Pipe Spring Ranch. In 1923 the ranch was purchased and set aside as a national monument, preserving the ranch house and two exterior buildings which were used as bunkhouses.

Vermillion Cliffs National Monument (El. 3,000' to 6,000')

Established in 2000 west of Glen Canyon Dam, the Vermillion Cliffs National Monument contains 293,000 acres. The cliffs are along the southern edge of the Paria Plateau, rising 3,000 feet in a spectacular escarpment. The monument also covers the Paria Plateau and the stunning Paria River Canyon.

The land was occupied by the Anasazi a thousand years ago, and was later traversed by Spanish explorers, Mexican traders and Mormon missionaries.

Desert bighorn sheep, pronghorn antelope, and mountain lions can be found on the monument land.

REFERENCES

PART I — THE LAND

THE COLORADO PLATEAU and THE BASIN AND RANGE PROVINCE

Billingsley, George H., Earle E. Spamer and Dave Menkes. "Mines and Prospectors of the Grand Canyon Region." *Quest for the Pillar of Gold*, Chpt. 3, Grand Canyon Association, Grand Canyon, AZ, 1997.

Chronic, Halka. *Roadside Geology of Arizona*. Mountain Press Publishing Co., Missoula, MT, 1986.

Fiero, Bill. *Geology of the Great Basin*. University of Nevada Press, Reno, NV, 1986.

Graf, W.L. (ed.), B. Hereford, J. Laity and R.A. Young. "Colorado Plateau." *Geomorphic Systems of North America*. Vol. 2, pps. 259-302, 1987.

Granger, Byrd H. *Arizona Place Names*. University of Arizona Press, Tucson, AZ, 1960.

Lucchitta, Ivo. "History of the Grand Canyon and of the Colorado River in Arizona." *Geologic Evolution of Arizona*. J.P. Penney and S.J. Reynolds, eds., Arizona Geological Society Digest, Vol. 17, Tucson, AZ, pps. 701-705, 1989.

Maley, Terry. *Exploring Idaho Geology*. Mineral Land Publications. Boise, ID, 1987.

Miller, Russel and the editors of Time-Life Books. *Planet Earth — Continents in Collision*. Time-Life Books, Alexandria, VA, 1983.

Priest, S.S., et al. *The San Francisco Volcanic Field, Arizona*. U.S.G.S. Fact Sheet, No. FS 0017-01, 2 pps., 2001.

THE COLORADO RIVER

Lingenfelter, Richard E. *Steamboats on the Colorado River, 1952-1916*. University of Arizona Press, Tucson, AZ, 1978.

Luchitta, Ivo. "History of the Grand Canyon and of the Colorado River in Arizona," *Geologic Evolution of Arizona*. J.P. Jenney and S.J. Reynolds, eds, Arizona Geological Society Digest, Vol. 17, Tucson, AZ, pps. 701-705, 1989.

Powell, J.W. *Exploration of the Colorado River of the West and Its Tributaries*, 1875.

Price, L. Greer. *An Introduction to Grand Canyon Geology*, Grand Canyon Association, Grand Canyon, AZ, 1999.

Wallace, Robert and the editors of Time-Life Books, *The Grand Canyon*. Time-Life Books, 1973.

PART II — THE PEOPLE

A CHRONICLE OF HISTORIC EVENTS IN NORTHWEST ARIZONA

"Western Migration Map Supplement." *National Geographic*, September 2000.

NATIVE AMERICANS IN NORTHWESTERN ARIZONA

Christensen, Don D. and Jerry Dickey. "Prehistoric Trail Associations: A Study in the Needles Region, Mojave Desert, California." *Rock Art Papers*, Ken Hedges (ed.), Vol. 13, San Diego Museum Papers 35, 1998.

Cordell, Linda S. *Prehistory of the Southwest*. Academic Press, Inc., New York, NY, 1984.

Cordell, Linda S. and George J. Gumerman. *Dynamics of Southwest Prehistory*. Smithsonian Institution Press, Washington, D.C., 1989.

Davis, Mary B. (ed.). *Native American in the Twentieth Century — An Encyclopedia*. Garland Publishing, Inc., New York, NY, 1996.

Dobyns, Henry F. "Hualapai Trails. A Report Submitted to Marks & Marks," Phoenix, AZ, December 20, 1954.

Reid, J. Jefferson and David E. Doyel. *Emil W. Haury's Prehistory of the American Southwest*. University of Arizona Press, Tucson, AZ, 1986.

Sturtevant, Wm. (general ed.) and Alfonso Ortiz (volume ed.). *Handbook of North American Indians: Volume 10 — Southwest*. Smithsonian Institution, Washington, D.C., 1983.

Waldman, Carl. *Atlas of the North American Indian*. Facts on File, Inc., New York, NY, 1985.

HISTORY OF MINING

Arizona Bureau of Mines, Bulletin 180. *Mineral and Water Resources of Arizona*. University of Arizona, Tucson, AZ.

Billingsley, George H., Earle E. Spamer and Dave Menkes. "Mines and Prospectors of the Grand Canyon Region," *Quest for the Pillar of Gold*, Chpt. 3, Grand Canyon Association, Grand Canyon, AZ, 1997.

Dewitt, Ed, Jon P. Thorson and Robert C. Smith. "Epithermal Gold Deposits; Part II. Geology and Gold Deposits of the Oatman District, Northwest Arizona." Chapter 1, *Geology and Resources of Gold in the United States*, U.S. Geological Survey Bulletin, 1857.

Durning, W.P. and L.J. Buchanan. *The Geology and Ore Deposits of Oatman, Arizona*. Arizona Geological Society Digest, Vol. 15, p. 141, 1984.

Koschmann, A.H. and M.H. Bergendahl. *Principal Gold-Producing Districts of the United States.* U.S.G.S. Professional Paper 610, 1968.

Paher, Stanley W. *Western Arizona Ghost Towns.* Nevada Publications, Las Vegas, NV, 1990.

Ransome, F.L. "Geology of the Oatman Gold District, Arizona — A Preliminary Report." U.S. Geological Survey Bulletin 743, 1923.

Schrader, F.C. *Mineral Deposits of the Cerbat Range, Black Mountains, and Grand Wash Cliffs, Mohave County, Arizona.* U.S. Geological Survey Bulletin 397, 1909.

"The Colorado River Region and John Wesley Powell." U.S. Geological Professional Paper, 669, 1969.

Wilson, Eldred D., J.B. Cunningham and G.M. Butler. *Arizona Lode Gold Mines and Gold Mining.* Arizona Bureau of Mines, Bulletin 137, Revised 1967.

THE RAILROAD: THE ATLANTIC & PACIFIC AND LATER, THE SANTA FE

Bryant, Keith L., Jr. *History of the Atchison, Topeka & Santa Fe Railway.* Railroads of America, MacMillan Publishing Co., New York, NY 1974.

Myrick, David F. *Railroads of Nevada and Eastern California.* Vol. 2, "The Southern Roads." Howell-North Books, Berkeley, CA, 1963.

Myrick, David F. *The Santa Fe Route, Railroads of Arizona,* Vol. 4, Signature Press, Welton, CA, 1998.

Weigle, Marty and Barbara A. Babcock (eds). *The Great Southwest of the Fred Harvey Company and the Santa Fe Railroad.* The Heard Museum, Phoenix, AZ, 1996.

MAN AND THE GRAND CANYON

Billingsley, George H., Earle E. Spamer and Dave Menkes. "Mines and Prospectors of the Grand Canyon Region." *Quest for the Pillar of Gold,* Chpt. 3, Grand Canyon Association, Grand Canyon, AZ, 1997.

Brown, Dick. *William Wallace Bass (1849-1933).* "The Ol' Pioneer Newsletter of the Grand Canyon Pioneer Society," 1999.

Dutton, C.E. "Tertiary History of the Grand Canyon District, with Atlas." U.S. Geological Survey. Monograph 2, 1882.

Grattan, Virginia L. and Mary Colter. "Boulder Upon the Red Earth." Grand Canyon Natural History Association, Grand Canyon, AZ, 1992.

Higgins, C.A. "Titan of Chasms." Information for Tourists, Passenger Department, The Santa Fe, Chicago, IL, 1908.

Powell, J.W. *Exploration of the Colorado River of the West and Its Tributaries,* Smithsonian Institute, 1875.

Powell, J.W. *Canyons of the Colorado,* 1895.

Rabbit, Mary C. *John Wesley Powell's Exploration of the Colorado River.* U.S. Department of the Interior/Geological Survey, 361-618/115, 1981

Suran, William. *The Kolb Brother's Biography with the Wings of an Angel — A Biography of Ellsworth and Emery Kolb, Photographers of the Grand Canyon.* http://kaibab.org/kolb, 1991.

Taylor, Karen L. *Grand Canyon's Long-Eared Taxi.* Grand Canyon Natural History Association. Grand Canyon, AZ, 1992.

U.S. Geological Survey. Monograph 2, 1882.

"The Colorado River Region and John Wesley Powell." U.S. Geological Survey Professional Paper 669.

Wallace, Robert and the editors of Time-Life Books. *The Grand Canyon.* Time-Life Books, 1973.

ROUTE 66

Casebier, Dennis G. *Camp Beale's Springs and the Hualapai Indians. Tales of the Mojave Road Number 7.* Tales of the Mojave Road Publishing Company, Norco, CA.

Casebier, Dennis. *Reopening the Mojave Road — A Personal Narrative. Tales of the Mojave Road Number 8.* Tales of the Mojave Road Publishing Company, Norco, CA, October 1983.

Tales of the Beale Road Publishing Company, 1985.

CIVILIAN CONSERVATION CORPS

Malach, Roman. "Home on the Range, Civilian Conservation Corps in Kingman Area," BLM Volunteer Program, Kingman Resource Area Office, Bureau of Land Management, 1984.

PART III — PLACES ON THE MAP

American Automobile Association. *AAA Road Atlas,* 1989.

Arizona Highways, "Mohave County, Arizona, U.S.A., Vol. 37, No. 5, May 1962.

Barnes, Will C. (revised and enlarged by Byrch H. Granger). *Arizona Place Names.* The University of Arizona Press, Tucson, AZ, 1960.

Goudy, Karin. "Life in a Boom Town — Oatman, Arizona." *History of Mining in Arizona.* Chapter 7, pps. 153-156, Michael J. Canty and Michael N. Greeley (eds.), Mining Club of the Southwest Foundation, Tucson, AZ, 1987.

Malach, Roman. *Oatman — Gold Mining Center.* Bicentennial Commemorative Publication, 1975.

Myrick, David F. *The Santa Fe Route, Railroads of Arizona,* Vol. 4, Signature Press, Welton, CA, 1998.

Schrader, F.C. "Mineral Deposits of the Cerbat Range, Black Mountains, and Grand Wash Cliffs, Mohave County, Arizona." U.S. Geological Survey Bulletin 397, 1909.

Rusho, W.L. and C. Gregory Crampton. *Lee's Ferry — Desert River Crossing,* Cricket Productions, Salt Lake City, UT, 168 pgs, 1992.

The World Book Encylopedia. Vol. 9. Field Enterprises Educational Corporation, Chicago, 1964.

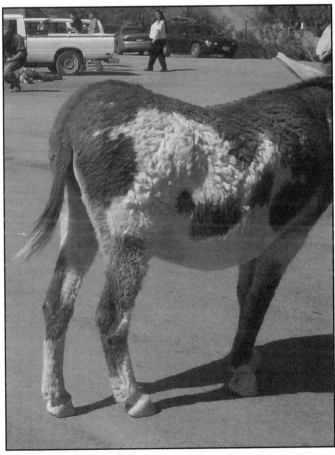

Oatman burro "On the Way Out."   Photo by Ann Ettinger.

"The End"